Fluid Power
Educational
Series

Electro-hydraulic Systems and Relay Circuits

Joji Parambath

Electro-hydraulic Systems and Relay Circuits

Copyright © 2020 Joji Parambath

All rights reserved

No part of this book may be reproduced or transmitted in any form or by any means, electronic or mechanical, including photocopying, recording, or by any information storage and retrieval system, without written permission from the publisher

ISBN: 9798654230027

https://jojibooks.com

Disclaimer of Liability

The contents of this textbook have been checked for accuracy. Since deviations cannot be avoided entirely, we cannot guarantee full agreement. Only qualified personnel should be allowed to install and work on hydraulic equipment. Qualified persons are those authorized to commission, to ground, and tag circuits, equipment, and systems following established safety practices and standards.

Dedicated to

Shri. Venkataiah, an outstanding staff member

Table of Contents

Chapter	Description	Page No.
--	Preface	vii
1	Introduction to Electro-hydraulic Systems	1
2	Solenoid Valves	6
3	Electrical Control Devices and Control Circuits	9
--	Objective Type Questions	36
--	Review Questions	36
Appendix 1	Graphic Symbols for Electrical Components	38
--	References	40

Preface

An electro-hydraulic system, in general, consists of an electrical or electronic control part controlling a hydraulic power part. Integrating the power density of hydraulic systems with the controlling possibilities of the electric systems opens up a new world of opportunities for the high-performing hydraulic power transmission systems. In this hybrid technology, solenoid valves are used as interfaces between the control part and the power part. The conventional solenoid valve acts as a converter that generates hydraulic outputs in response to electrical input signals. Control and feedback elements like pushbuttons (PBs), relays, sensors, and timers are used in the electro-hydraulic systems.

Many other fluid power topics are given in other textbooks under the fluid power educational series by the same author. A list of all the textbooks is given at the end of the book (Page No. 41). Also, please see the details at https://jojibooks.com

Enjoy reading the book.
Your feedback is most welcome

JOJI Parambath

About the Author

Joji Parambath is a trainer in the field of Pneumatics, Hydraulics, and PLC, for over 25 years. During his career, he has trained numerous professionals from the industries as well as faculty members and students of engineering institutions.

At present, he is the key trainer at Fluidsys Training Center, Bangalore, India, (https://fluidsys.org) which is providing training in the field of Pneumatics and Hydraulics. He has already written two books on Pneumatics and Hydraulics. The publication of the present series of 32 books is intended to restructure and update the existing books.

The author wishes to thank all trainees for their lively interaction and many useful suggestions during the training programmes that prompted the author to write the present series of books. You may send your feedback to joji.p@hotmail.com

10th June 2020

Chapter 1 | Introduction to Electro-hydraulic Systems

Hydraulic valves, in general, can be actuated manually or electrically or by other means. In the electrical actuation, the necessary actuating force is obtained electrically with the help of the essential actuating element called 'solenoid'. A hydraulic valve with a solenoid is the well-known electro-hydraulic solenoid valve.

An electro-hydraulic system consists of a hydraulic power transmission system and an electrical control system. A solenoid valve in an electro-hydraulic system acts as an interface between the hydraulic power system and the electrical control system. With the use of the solenoid valve, the advantages of both the hydraulic and electrical media can be exploited. The result is that electro-hydraulic systems are extensively used in many industrial machines and production systems.

The electro-hydraulic valves are usually classified into three fundamental categories: (1) discrete valves, (2) proportional valves, and (3) servo-valves. Conventional discrete (on/off) solenoid valves and various control circuits using them are the subject matter of this book. The details of proportional valves and servo valves are described in other books by the same author.

Hydraulic Power Transmission System

The block diagram of a hydraulic power system is given in Figure 1.1. Pressurized oil medium is generated by a pump driven by a prime mover, such as a motor. The energy in the form of pressurized oil is then transmitted to some actuators, such as cylinders, and hydraulic motors, through final control elements, such as directional control valves. A relay controller or PLC controller controls the final control elements.

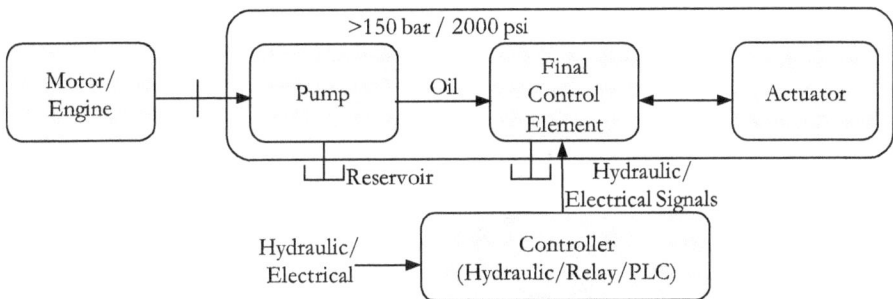

Figure 1.1 | Block diagram of a hydraulic power system

Hydraulics is an engineering science of liquid pressure and flow. Hydraulic systems are high-pressure systems. They are suitable for systems that require precise slow speed control or involve the holding of heavy loads.

The following sections present the details of hydraulic power packs and hydraulic actuators, briefly. The details of solenoid valves are given in Chapter 2.

Hydraulic Power Packs

A hydraulic power pack, as shown in Figure 1.2, is a unit that supplies the required fluid to the system. Therefore, a reservoir in the power pack should maintain a sufficient amount of high-quality fluid at all times for its efficient operation.

A power pack is a compact, portable, and pre-configured assembly consisting of some essential components and some optional components. The essential components are a fluid-filled reservoir, close-coupled pump-motor unit, pressure relief valve, and pressure gauge, and the optional components include a heat exchanger, temperature controller, directional control valves, and filters.

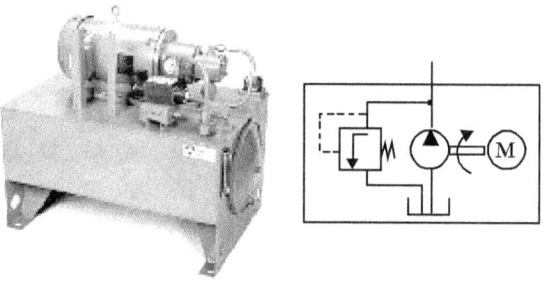

Figure 1.2 | A hydraulic power pack
Courtesy: Advance Motion Control, USA

Hydraulic Pumps

A hydraulic pump is a positive displacement device that converts mechanical power into hydraulic power. Hydraulic pumps can be classified according to the types of pumping elements, such as gears, vanes, and pistons, used. Accordingly, there are three main types of pumps used in hydraulic systems. They are: (1) Gear pumps, (2) Vane pumps and (3) Piston pumps.

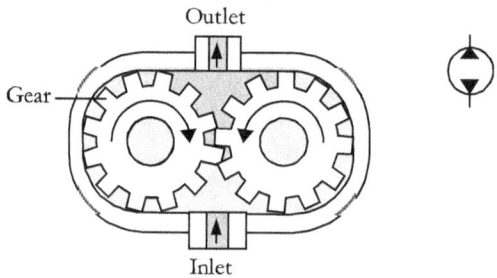

Figure 1.3 | A gear-type pump

Figure 1.3 gives the cross-sectional view of a gear pump. It consists of two close-meshing identical gears, enclosed in a close-fitting housing. Further, the gears rotate in opposite directions when driven by a prime mover. When the gears rotate, the diverging teeth create an expanding volume at the inlet side of the pump. This expanding volume creates a partial vacuum at the inlet of the pump, which draws fluid into the chamber from the associated reservoir. The fluid is trapped in the chambers on either side. The trapped fluid then moves around the periphery of the rotating gears as two streams. Remember, the pump has a positive internal seal against leakage. The two streams, then, recombine and are positively ejected out of its delivery port. Therefore, when run by the drive motor, the gears displace a fixed volume of the fluid per revolution of its driveshaft and create a flow.

Pressure Relief Valve (PRV)

A PRV, as shown in Figure 1.4, regulates the pressure in a hydraulic system. It consists of a body with an inlet port (P) and an outlet (tank) port (T). It also consists of a poppet that remains pressed against the valve seat by a heavy-duty spring. The valve is, usually, closed with the spring bias. In most of these valves, an adjusting screw is provided to vary the spring tension externally and hence to set the system pressure.

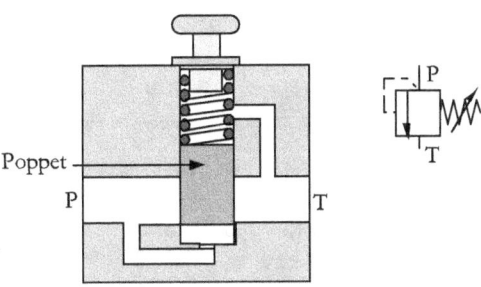

Figure 1.4 | A pressure relief valve (PRV)

The system pressure acts directly on the spring-biased poppet within the PRV. When the system pressure is below the PRV's set pressure, the flow through the valve is blocked. When the system pressure exceeds the preset pressure, the poppet is lifted off its seat. This action makes an auxiliary passage open and diverts the fluid away from the pressurised section of the system to the system reservoir. When the excess pressure is relieved, the auxiliary passage closes again. In this way, the system pressure remains on the value as determined by the setting of the PRV.

Hydraulic Actuators

A hydraulic actuator is a positive displacement device used in a hydraulic system to drive the attached load to get some useful work in the system. Its primary function is to convert hydraulic power into mechanical power. The resulting output motion can be either linear or rotary. Accordingly, there are two basic types of hydraulic actuators. They are: (1) Linear actuators and (2) Rotary actuators.

The linear actuators convert hydraulic energy into straight-line mechanical energy, and the rotary actuators convert hydraulic energy into rotary mechanical energy. An example of the linear actuator is the cylinder, and the rotary actuator is the hydraulic motor.

Linear Actuators

A linear hydraulic actuator converts hydraulic power into a controllable linear force or motion or both. A cylinder is an example of a linear actuator. Hydraulic cylinders can be classified into the following types: (1) Single-acting cylinders and (2) Double-acting cylinders.

Single-acting Cylinders

Figure 1.5 shows the cross-sectional view of a single-acting cylinder. It consists of a barrel, a piston-and-rod assembly, a spring, end-caps, seals, and a port.

A fluid chamber is formed in the cylinder with the barrel, piston, and cap-end endplate. The piston-and-rod assembly is a tight fit inside the barrel and is biased by the spring. The port is integrated into its cap-end, and that admits or relieves the system fluid.

The application of pressure through the port moves the piston-and-piston-rod assembly in one direction to provide the working stroke. The piston-and-piston-rod assembly moves in the opposite direction, either by spring force or gravity. In the cylinder with a spring-assisted retraction, the spring is designed not to carry any load, but, to retract the piston-and-piston-rod assembly with sufficient speed. The single-acting cylinder is capable of performing work only in one direction of its motion and hence the name 'single-acting cylinder'.

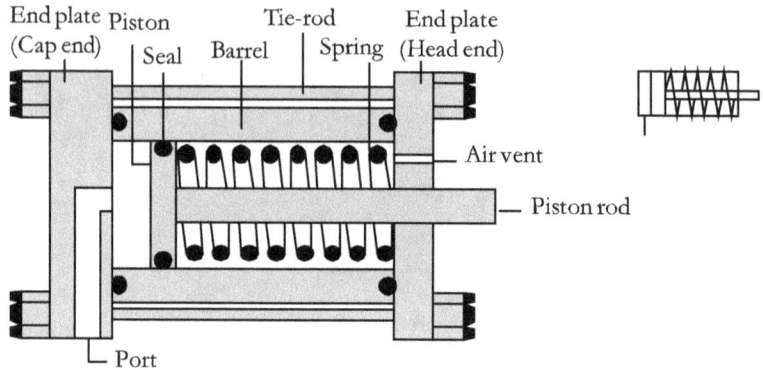

Figure 1.5 | A single-acting hydraulic cylinder

Double-acting Cylinders

Figure 1.6 gives the cross-sectional view of a double-acting hydraulic cylinder. It consists of a barrel, piston-and-piston-rod assembly, end-caps, seals, and two ports. Further, the double-acting cylinder has fluid ports on both ends, namely the piston-side port, and the piston-rod-side port. The application of pressure through the piston-side port extends the cylinder, provided that the pressure from the piston-rod side is relieved.

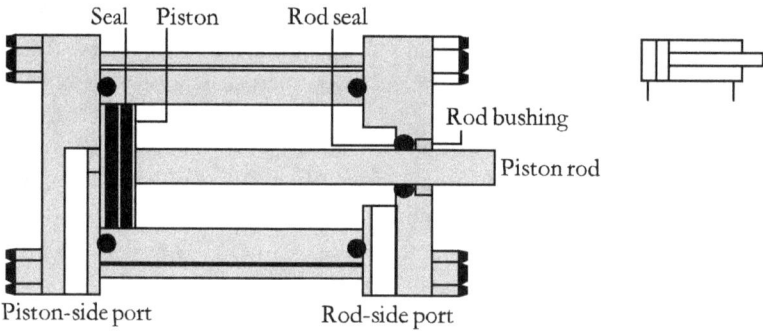

Figure 1.6 | A double-acting cylinder

In the same way, the application of pressure to the piston-rod-side port retracts the cylinder, provided that the pressure from the piston side is relieved. A double-acting cylinder can perform work in both directions of its motion, and hence the name 'double-acting cylinder'.

Hydraulic Rotary Actuators

Hydraulic rotary actuators are the muscle behind the rotary motions in industrial and mobile hydraulic systems. They are positive-displacement devices that convert hydraulic energy to rotary mechanical energy when supplied with flowing fluid under pressure. A rotary actuator converts the system pressure and flow, to a controllable rotary force (torque) or rotary motion or both. Here, the fluid pressure is converted to the torque, and the flow is converted to the rotary speed. The torque and motion of the rotary actuator can be used for obtaining the rotary operation in industrial machinery.

Hydraulic rotary actuators can be classified into two types. They are: (1) Semi-rotary actuators and (2) Motors. A semi-rotary actuator (or oscillating motor) is capable of producing only limited rotation and can twist objects along a partial arc. On the other hand, a hydraulic motor is capable of producing continuous rotation and can impart continuous rotary motion to the connected load.

Hydraulic Motors

A hydraulic motor mainly consists of a set of moving elements, such as gears, vanes, or pistons connected to the output shaft of the motor, and enclosed in a single housing. Figure 1.7 shows the schematic diagram of a gear motor. The shaft rotates when the pressurized system fluid is applied to the intermeshing gears. In this way, the motor is capable of converting the applied pressure to rotary mechanical force and consequently driving the load attached to the motor. The fluid returns to the system reservoir, after passing through the motor.

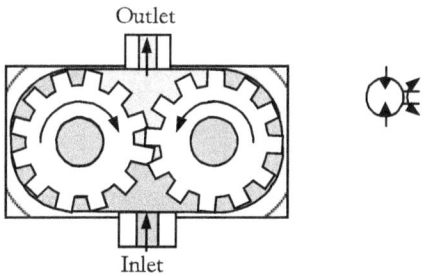

Figure 1.7 | Symbolic diagrams of hydraulic motors

Electro-Hydraulic Systems

In electro-hydraulics, a hydraulic power transmission system is controlled by an electrical control system. This technique permits the optimum utilization of different types of energy media for cost-effective and efficient production systems. Next, a final control element, such as a solenoid valve, acts as the interfacing component between the hydraulic power system and the electrical control system. The control system mainly consists of an electromagnetic relay controller and many input devices such as pushbuttons and sensors. In the relay controller, many relays are interconnected, as per the governing logic, to control the final control elements (solenoid valves) and hence to achieve the desired control function. In complex applications, electrical controls are employed almost exclusively.

This book explains the working of solenoid valves and various electrical control components, such as pushbuttons (PBs), relays, sensors, timers, pressure switches, and counters. The book also presents the details of a variety of sensors and the development of many electro-hydraulic circuits, including circuits using sensors for the automatic operation of electro-hydraulic systems. The objective of this book is to help the reader gain a thorough understanding of relay circuits and their automation with the help of typical examples. The relay circuits provided in this book generally increase in complexity, and the reader is encouraged to work through them in the sequence given in this book.

Chapter 2 | Solenoid Valves

The magnetic effect of electric current can be used for implementing various technical functions in industrial systems. When an electric current passes through a straight wire, magnetic lines of force are produced around the wire in a manner, as shown in Figure 2.1(a). The direction of these lines of force depends on the direction of the current flow. However, this force is distributed over the length of the wire. Hence, it cannot be used for realizing any useful control function.

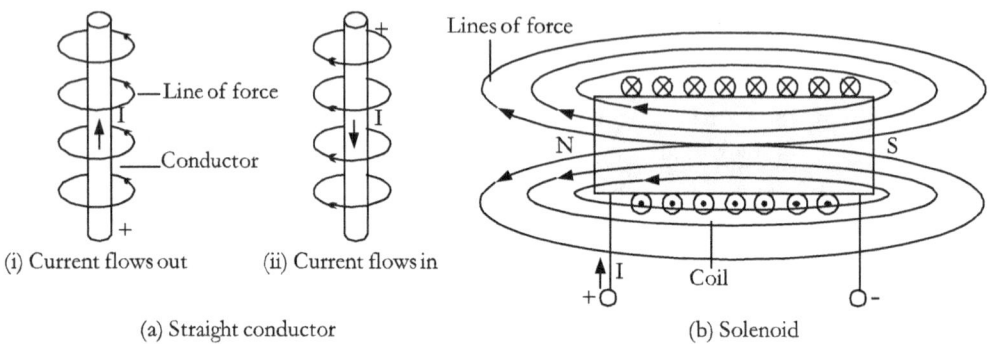

Figure 2.1 | A graphic representation to show the magnetic effect of electric current

Therefore, to concentrate the magnetic lines of force, the wire is wound in the shape of a long cylindrical coil. This formation of the coil, as shown in Figure 2.1(b), is usually known as the solenoid. A soft iron core introduced into the solenoid strengthens the magnetic force further. However, another useful property of the solenoid coil with an off-center core is used to develop the actuating force in the valve. That is; the core of the coil is pulled towards the center of the coil when the electric current is passed through it. The resulting linear force can be used for actuating the valve.

AC solenoids Vs DC Solenoids
In general, solenoid coils can be designed to operate either on AC or DC power. Many differences can be recognized between AC and DC solenoids. An AC solenoid develops a high initial inrush current of about five times more than its rated current. If the armature in the AC solenoid is not allowed to shift entirely to its end position, the current drawn by it remains high. The result is that the coil overheats and burns out. On the other hand, a DC solenoid does not experience much inrush current and hence do not run the risk of being damaged by the initial current. Therefore, the armature of the DC solenoid can remain partially shifted to a particular position indefinitely without an increase in the current drawn by it.

Solenoid Valves, Discrete
A discrete electro-hydraulic solenoid valve consists of a hydraulic section and an electrical control section. The hydraulic section includes a valve body and a spool for the direction control and the electrical part consisting of a solenoid and a core (plunger) for linking and controlling the hydraulic section. The core is usually positioned away from the center of the coil by a biasing spring. When we energize the coil, the resultant magnetic field pulls the core towards the center of the coil. This discrete movement of the core is used to actuate the solenoid valve. Thus, the incoming electrical signal at the input of the valve can be converted to corresponding fluid flow at the output of the valve.

Discrete electro-hydraulic solenoid valves can be classified as the direct-acting type and the pilot-operated type. In the direct-acting type solenoid valve, an orifice in the valve is opened directly by the solenoid and its plunger. The pilot-operated solenoid valve consists of an internal pilot valve and the main valve. In this type of valve, a pilot electrical signal controls the opening of the pilot valve, which in turn provides the actuating force for the main valve.

Further, the discrete electro-hydraulic solenoid valves can be classified according to the type of port/position configuration, such as 3/2-way, 4/2-way, and 4/3-way directional control valves. The following sections present the principles of operation of some typical discrete solenoid valves.

3/2-way Single-solenoid Valve, Spring Return

Figure 2.2 shows the cross-sectional views of a 3/2-way single-solenoid electro-hydraulic valve in its normal and actuated positions. It consists of an electrical solenoid part controlling a hydraulic power part. Next, the electrical part consists of a solenoid (Y) with an armature. The 3/2-way valve consists of a body with a spool inside, a reset spring, and necessary ports (P, A, and T, with usual notations).

In the normal position of the solenoid valve, the pressure port P remains blocked, and the working port A is connected to the tank port T internally. When the rated voltage is applied to the coil Y, the armature is forced towards the center of the coil. The movement of the armature shifts the spool away from the valve seat. The valve is said to be in its actuated position. In the actuated position of the valve, the port P is connected to the port A, and the port T remains blocked. When the supply to the coil is cut off, the valve returns to its normal position. This valve can be used as the final control element for controlling a single-acting cylinder.

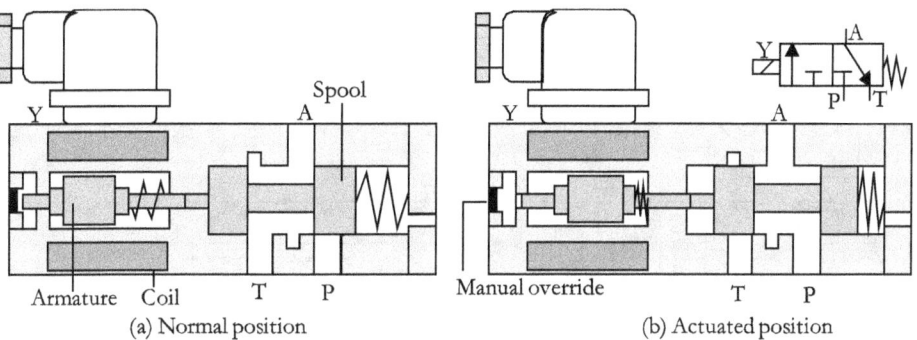

Figure 2.2 | Cross-sectional views of a 3/2-way single-solenoid hydraulic valve

The manual override facility is provided on the solenoid valve as a standard feature. This feature can be used to operate the valve manually by turning an eccentric screw provided for the purpose. Further, a light-emitting diode (LED) can be incorporated into the plug housing of the solenoid valve for the visual indication of the ON/OFF state of the coil.

4/2-way Single-solenoid Valve, Spring Return

Figure 2.3 shows the cross-sectional views of a 4/2-way single-solenoid electro-hydraulic valve in its normal and actuated positions. It consists of an electrical part controlling a hydraulic power part. Further, the electrical part consists of a solenoid with an armature. The 4/2-way hydraulic valve consists of a body with a spool inside, a reset spring, and necessary ports (P, A, B, and T, with usual notations).

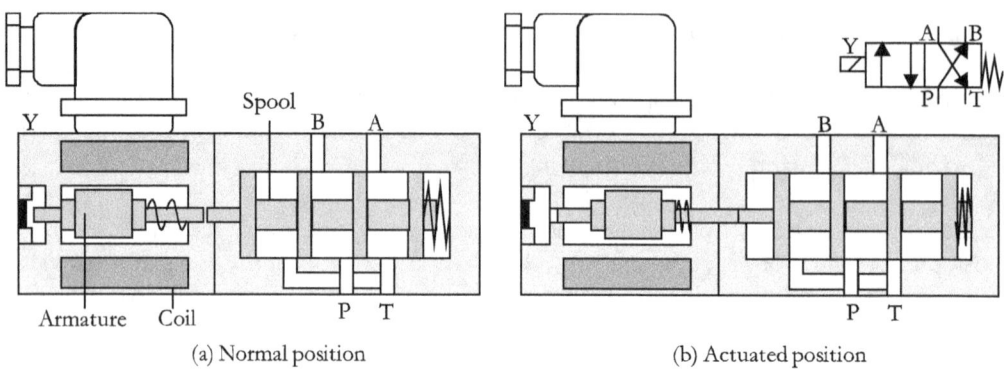

Figure 2.3 | Cross-sectional views of a 4/2-way single solenoid valve

In the normal position of the solenoid valve, the pressure port P is connected to the working port B, and the working port A is connected to the tank port T. The valve is actuated when the rated voltage is applied to the coil Y. In the actuated position, the port P is connected to the port A, and the port B is connected to the port T. When supply to the coil is cut off, the valve returns to its normal position. This valve can be used as the final control element for controlling a double-acting cylinder.

4/2-way Double-solenoid Valve

Figure 2.4 shows the cross-sectional view of a 4/2-way double-solenoid electro-hydraulic valve. It consists of two solenoid coils (Y1 and Y2) on either side of the valve controlling its hydraulic part. The 4/2-way double-solenoid hydraulic valve consists of a body with a spool and necessary ports (P, A, B, and T, with usual notations). This type of valve design is noted for the absence of a reset spring.

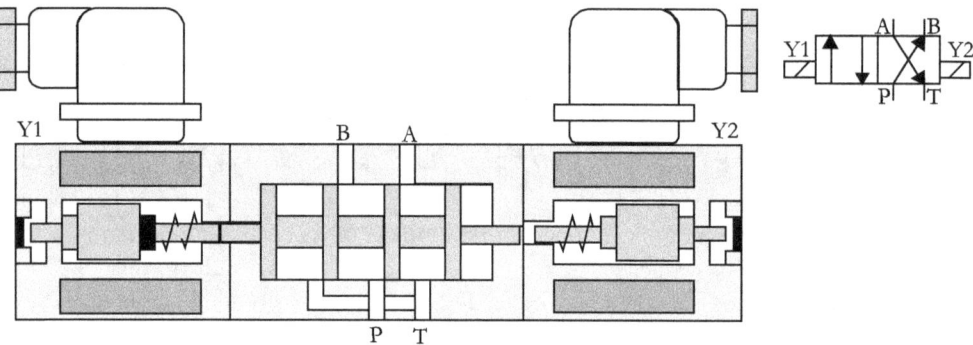

Figure 2.4 | A cross-sectional view of a 4/2-way double solenoid valve in its normal position

When the rated voltage is applied to the coil Y1 momentary or continuously, the valve is actuated to a particular switching position. In this switching position, the pressure port P is connected to the working port A, and the working port B is connected to the tank port T. This position is maintained by the valve until a signal is applied to the coil Y2. When the rated voltage is applied to the coil Y2 momentary or continuously, the valve is actuated to the other switching position. In this switching position, the pressure port P is connected to the working port B, and the working port A is connected to the port T. This position is maintained by the valve until a signal is applied to the other coil Y1.

Chapter 3 | Electrical Control Devices and Control Circuits

Many a time, a technician is overwhelmed by the complexity and size of control systems for industrial applications like automatic manufacturing assembly lines. However, this complexity is no issue for those who understand the fundamentals of control and the principle of operation of essential control components. Remember, even a complex control system is made up of a series of simple control circuits involving basic control components. Some basic control components are: pushbuttons, limit switches, float switches, pressure switches, flow-switches, thermostats, relays, proximity sensors, timers, and counters. The functional aspects of many control components and some typical control circuits are presented in the following sections. A list of important graphic symbols used for electrical components is given in Appendix 1.

Switch

An electrical switch is a device consisting mainly of a set of control contacts for making or breaking an electrical circuit. In control applications, switches are integrated as control contacts in various pilot devices such as pushbuttons, limit switches, float switches, pressure switches, timers, and counters. The purpose of control contacts is to present electrical signals from various points in the control system to the area of signal processing.

Pushbuttons

The cross-sections of various types of pushbuttons in the normal and actuated positions and their symbols are given in Figure 3.1.

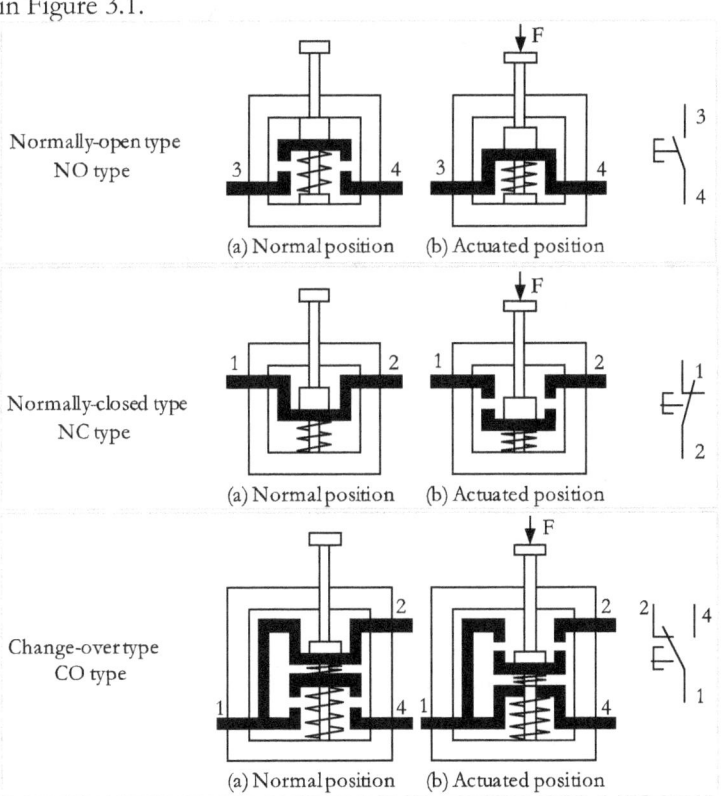

Figure 3.1 | Types of pushbuttons and their symbols

A pushbutton is a switch used to close or open an electric control circuit. This device consists of a set of fixed and movable contacts and a restraining spring. Pressing the pushbutton against the restraining spring operates its contacts. Pushbuttons are of two types: (i) Momentary-contact type, and (ii) Maintained-contact type (or detent type). In the momentary-contact type, the contacts are operated only when the pushbutton is pressed continuously, and the contacts return to their normal position when the pushbutton is released. In the maintained-contact type, the contacts are operated when the pushbutton is pressed and remain in that position even if the actuating force is removed. The contacts return to the original position only when the pushbutton is pressed again.

The contacts of the pushbuttons, distinguished according to their functions, are as follows: (i) Normally Open (NO) type, (2) Normally Closed (NC) type and (3) Change-Over (CO) type. In the NO type, the contacts are open in the normal position, inhibiting the energy flow through them; and in the actuated position, the contacts are closed, permitting the energy flow. In the NC type, the contacts are closed in the normal position, permitting the energy flow through them; and in the actuated position, the contacts are open, inhibiting the energy flow. Changeover contact is a combination of NO and NC contacts.

Terminal Markings of Contacts

Contacts are used in many types of pilot devices such as pushbuttons, relays, timers, and counters. To identify the terminals of contact, they are designated with a set of numbers based on the function of the contact. The numbering system for the contact terminals is given in Table 3.1

Table 3.1 | Terminal markings of electric contacts

Type of pilot device	Terminal numbers of	
	NC contact	NO contact
Ordinary devices (PBs, relays)	1 and 2	3 and 4
Special devices (Timers, counters)	5 and 6	7 and 8

Pushbutton Station

A pushbutton station consists of many contacts (NO, NC or CO) pairs/sets with a common actuation, as shown in Figure 3.2(a). This device is compact and less expensive but usually comes with limited current carrying capacity.

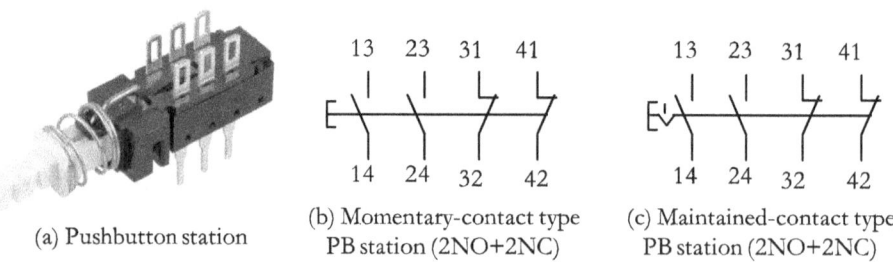

(a) Pushbutton station (b) Momentary-contact type PB station (2NO+2NC) (c) Maintained-contact type PB station (2NO+2NC)

Figure 3.2 | Symbolic representations of Pushbutton stations.

Figure 3.2(b) shows the symbol of a momentary-contact type pushbutton station with a contact configuration of 2 NO + 2 NC. Figure 3.2(c) shows the symbol of a maintained-contact type pushbutton station. A pair of consecutive two-digit numbers are used to designate the terminals of a contact in the pushbutton station. In the two-digit number, the unit place indicates the function of the contact (that is, whether it is a NO or NC type). The digit at the ten's place merely represents a serial ordering of all contact pairs in the pushbutton station for identifying each contact pair uniquely. In pushbutton stations, the required actuating force increases as the number of contacts increases. For this reason, the number of contacts is usually limited in pushbutton stations.

Industrial Control Voltages

In earlier days, the control voltages used in industries were 230 V AC, 110 V AC, etc. However, the tendency was to reduce the control voltage to a lower level from the operator's safety point of view. At present, 24 V DC is the standard industrial control voltage in many countries. The possible voltage levels for electrical control components, used in industrial systems are: 12 V DC, 24 V DC, 24 V 50/60 Hz, 48 V 50/60 Hz, 110/120 V 50/60 Hz, and 220/230 V 50/60 Hz. A DC power pack is typically used to convert 230 V AC supply to 24 V DC supply. In this textbook, all the electro-hydraulic control circuits are shown with 24 V DC control voltage.

Example 3.1 | Direct control of a single-acting cylinder

A small-volume single-acting cylinder used in a hydraulic system with a fixed-displacement pump is to extend when a pushbutton (PB) is pressed. The cylinder is to retract when the pushbutton is released. A 3/2-way single solenoid valve, rated for a coil voltage of 24 V DC, controls the cylinder. Develop a hydraulic power circuit and an electrical control circuit to implement the control task.

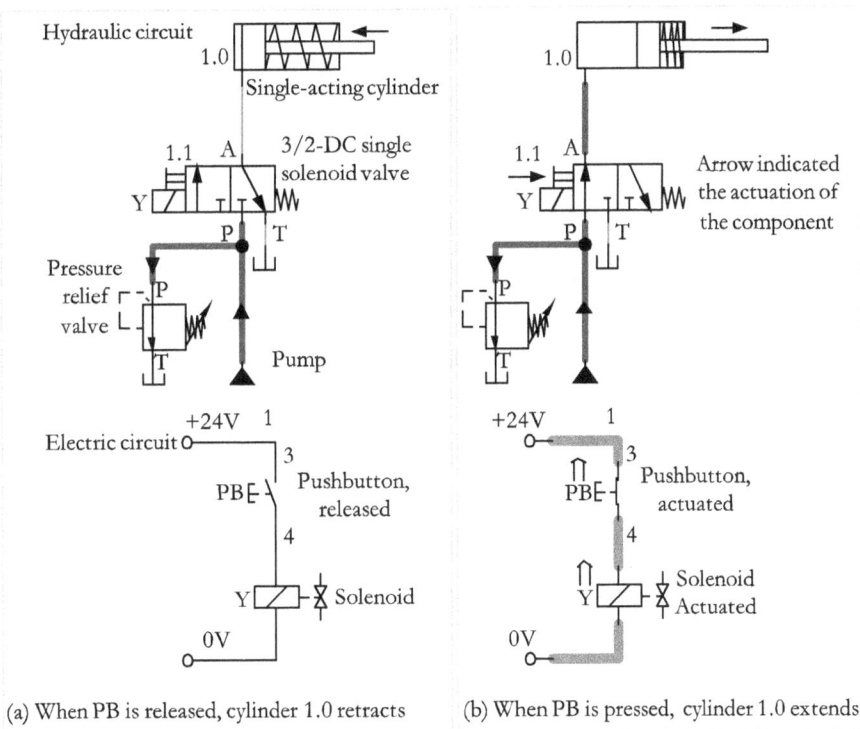

(a) When PB is released, cylinder 1.0 retracts   (b) When PB is pressed, cylinder 1.0 extends

Figure 3.3 | Two circuit positions for the direct control of a single-acting cylinder (Example 3.1)

Solution

An electro-hydraulic circuit diagram is conventionally drawn with two distinct parts. First, the hydraulic power circuit is drawn, and then the electrical control circuit is drawn just below the hydraulic circuit. The interface between the hydraulic and electrical elements is the solenoid coil Y that appears on the hydraulic circuit and electrical circuit with a common designation. The pressure port P is connected to the power supply, and the tank port T is connected to the reservoir.

The desired control task by the single-acting cylinder is illustrated with two positions of the circuit in Figure 3.3. The actuation of the pushbutton (PB) generates current flow through solenoid coil Y, which in turn causes the actuation of the 3/2-DC single solenoid valve, as shown in Figure 3.3(b). The fluid then flows from pressure port P to the cylinder through the working port A of the valve, and the cylinder extends.

When the PB is released, the electrical circuit is interrupted. The solenoid coil is de-energized, and the valve returns to its original position, as shown in Figure 3.3(a). In this position, the pressure port P is blocked, and the working port A is connected to the reservoir through the tank port T. The fluid contained in the cylinder is discharged to the reservoir, and the cylinder retracts.

Example 3.2 | Direct control of a double-acting hydraulic cylinder

A double-acting cylinder used in a hydraulic system is to extend when a pushbutton (PB) is pressed and to retract when the PB is released. The system uses a fixed-displacement pump. A 4/2-way single-solenoid electro-hydraulic DC valve controls the cylinder. Develop the hydraulic power circuit and the electrical control circuit to implement the control task.

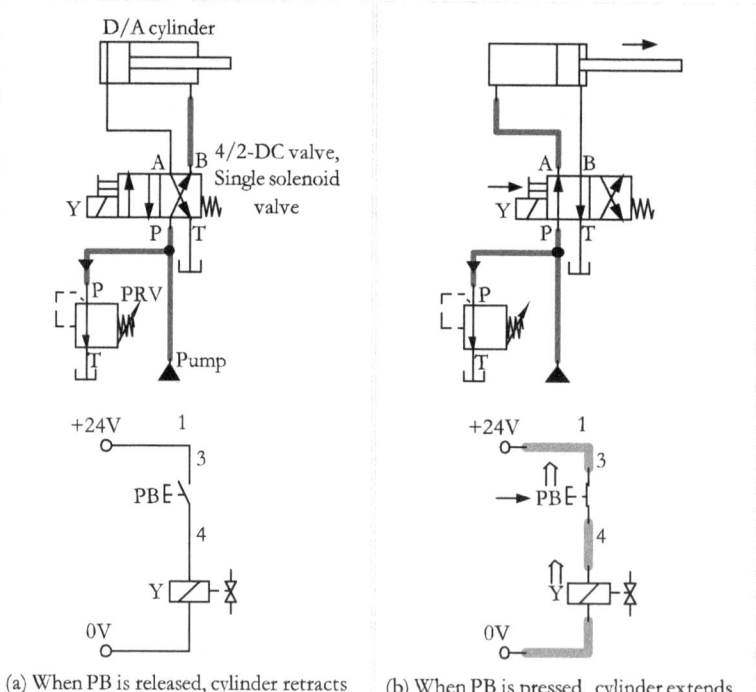

(a) When PB is released, cylinder retracts (b) When PB is pressed, cylinder extends

Figure 3.4 | Two positions of an electro-hydraulic circuit for the direct control of a double-acting hydraulic cylinder. (Example 3.2)

Solution

Figure 3.4 illustrates two positions of the electro-hydraulic circuit for the direct control of the double-acting cylinder using the electro-hydraulic solenoid valve with a pressure port 'P', working ports 'A' and 'B', and a tank port 'T'. The pressure port P is connected to the power supply, and the tank port T is connected to the reservoir. A power pack with the fixed-displacement pump supplies the necessary hydraulic power.

The pushbutton (PB) is used to control the solenoid. The actuation of the PB generates a current flow through the solenoid coil Y and actuates the valve, as shown in Figure 3.4(b). The fluid then flows from the port P to the port A and acts on the piston side of the cylinder. The return fluid flows through the valve from the port B to the port T. The cylinder then extends to its forward end position.

When the PB is released, the electrical circuit is interrupted, as shown in Figure 3.4(a). The solenoid coil is de-energized, and the valve returns to its original position. The fluid then flows from the port P to the port B and acts on the piston-rod side of the cylinder. The return fluid flows through the valve from the port A through the port T to the system reservoir. The cylinder then retracts to its home position.

Electromagnetic Relay

Figure 3.5 shows an electromagnetic relay. It is an electromagnetically actuated switch that operates under the control of an additional electrical circuit.

The relay mainly consists of a coil and many contact sets. Further, each contact set consists of a stationary contact and a movable contact. It also includes a stationary core and a movable core to confine the magnetic field. The movable contacts are coupled to the movable core.

Therefore, when the coil is energized, the movable core is pulled towards the stationary core, thus operating all its coupled contacts simultaneously. This movement either makes or breaks the connection of the movable contact with its respective fixed contact in each contact set.

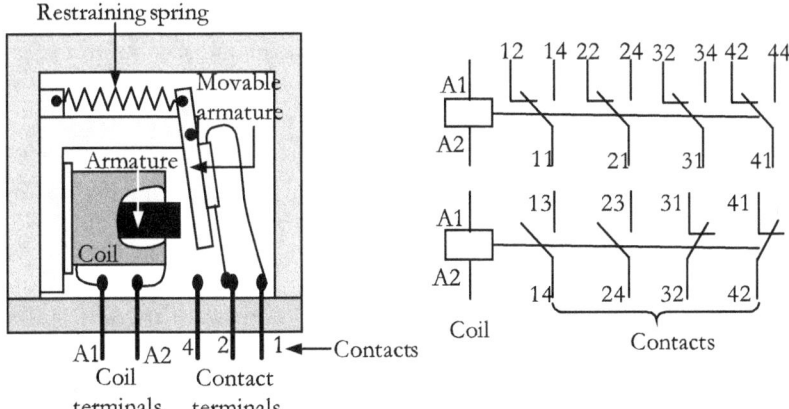

Figure 3.5 | A cross-sectional view of an electromagnetic relay

Example 3.3 | Indirect control of a double-acting hydraulic cylinder using a relay.

A large-volume double-acting cylinder used in a hydraulic application is to be controlled by using a 4/2-way single-solenoid electro-hydraulic DC valve. The application uses a fixed-displacement pump. The cylinder is to extend when a pushbutton (PB) is pressed and to retract when the PB is released. Develop the electro-hydraulic control circuit to implement the control function.

Solution

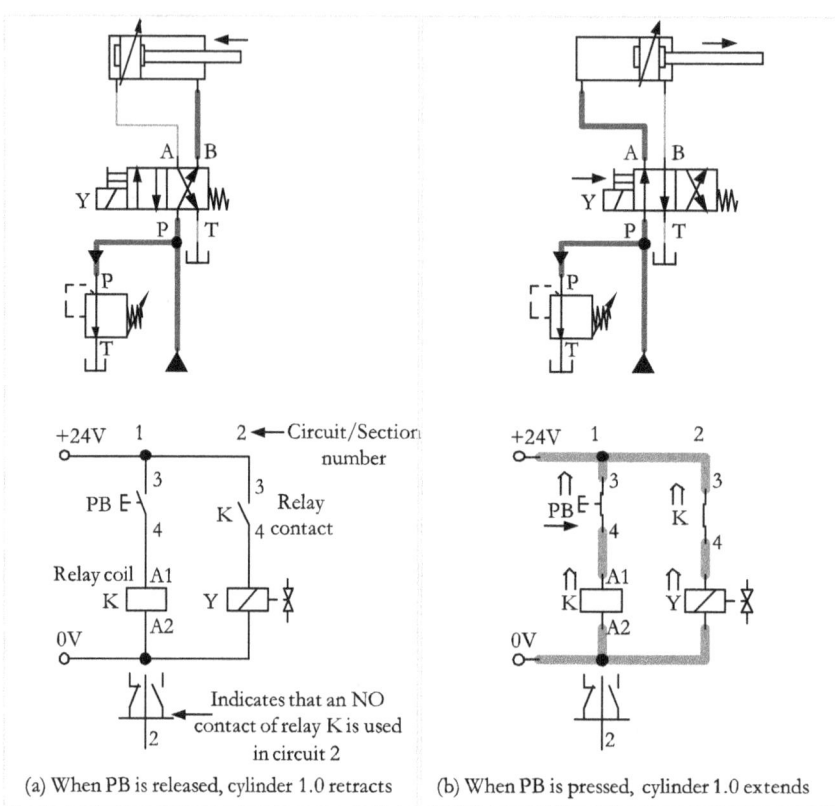

(a) When PB is released, cylinder 1.0 retracts   (b) When PB is pressed, cylinder 1.0 extends

Figure 3.6 | Two positions of an electro-hydraulic circuit for the indirect control of the double-acting hydraulic cylinder using a relay. (Example 3.3)

Figure 3.6 illustrates two positions of the electro-hydraulic circuit for the indirect control of the double-acting hydraulic cylinder with the 4/2-way single-solenoid valve with a pressure port 'P', working ports 'A' and 'B', and a tank port 'T', through a relay (K). A power pack with the fixed-displacement pump supplies the necessary hydraulic power. The pushbutton (PB) is used to control the relay coil. The actuation of the PB energizes the coil, and consequently, all of its contacts are operated. The solenoid coil Y is energized through the NO contact of the relay in branch 2, causing the actuation of the solenoid valve, as illustrated in Figure 3.6(a). The fluid then flows from the port P to the port A of the valve and acts the piston side of the cylinder. The return fluid flows through the valve from the port B to the port T. The cylinder then extends to its forward end position.

When the PB is released, the electrical circuit in branch 1 is interrupted. Both the relay coil and the solenoid coil are de-energized, and the valve returns to its original position, as illustrated in Figure

3.6(b). The fluid then flows from the port P to the port B and acts on the piston-rod side of the cylinder. The return fluid flows through the valve from the port A through the port T to the system reservoir. The cylinder then retracts to the rear end position.

It is usual to show the consolidated information on the connection positions of the relay contacts in the circuit diagram just below the symbol of the relay coil. For example, the use of the NO contact of the relay K in branch 2 of the circuit is indicated below the coil, as shown in Figure 3.6.

Logic Controls

Logic circuits are designed to carry out decision-making output functions based on many input signals from pushbuttons or sensors representing certain conditions of the associated machines or systems. Two of the essential basic logic functions are: 'AND' and 'OR'. Signal levels in logic devices are characterised by the following two states: logic '1' and logic '0'. Usually, the logic '1' represents an ON state, and the logic '0' represents an OFF state. The block diagrams and truth tables for the OR logic and AND logic functions are given in Figure 3.7(a) and (b), respectively.

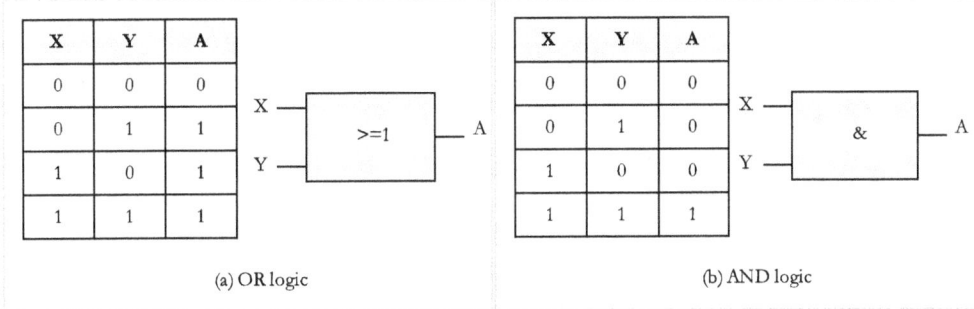

Figure 3.7 | Block diagrams and truth tables for 'OR' and 'AND' logic functions

In the OR logic function, there is an output when one or more inputs are present. An example of the OR logic function is a lamp that is controlled from one or more positions. A parallel connection of input devices makes an OR function. Its operation is illustrated in the self-explanatory Figure 3.8(a) for the control of a lamp using two pushbuttons connected in parallel.

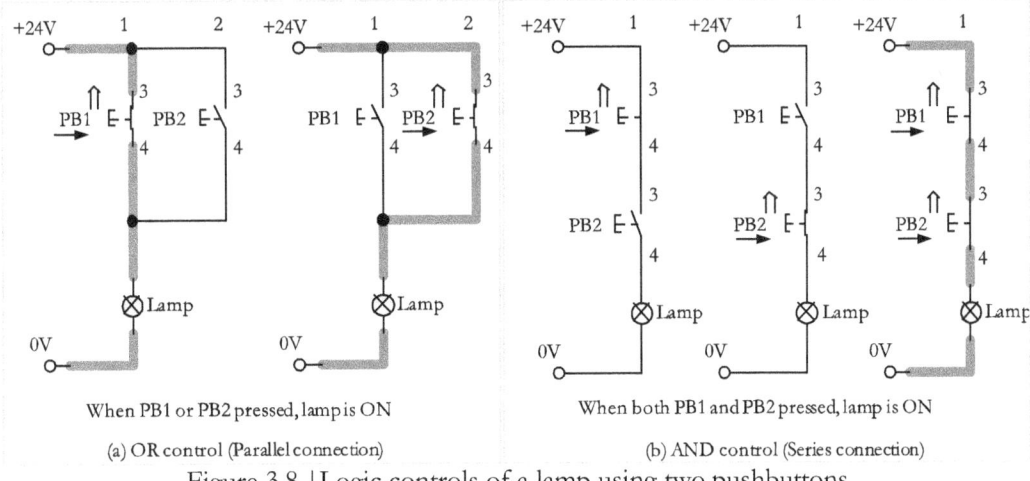

Figure 3.8 | Logic controls of a lamp using two pushbuttons

In the AND logic function, there is an output only when all the inputs are present. An example of the AND logic function is a cylinder that extends only when its safety guard is placed in position, and a 'start' signal is given. A series connection of input devices, like pushbuttons and sensors, gives an AND function. The control of a lamp using two pushbuttons connected in series for the AND logic function is illustrated in Figure 3.8(b). Any complex logic function can also be set up easily by a series-parallel connection of input devices.

Example 3.4 | Control of a double-acting cylinder from two locations

A large-volume double-acting hydraulic cylinder is to be controlled using two pushbuttons (PB1 and PB2) installed at two different locations. The cylinder should extend when either PB1 or PB2 is pressed. The cylinder should retract and remain in the home position when PB1 and PB2 are released. Develop an electro-hydraulic circuit to implement the control task.

Solution

The electro-hydraulic circuit in three critical positions is given in Figure 3.9. As shown in the hydraulic part of the circuit, the double-acting cylinder is controlled by a 4/2-way single solenoid spring-return valve. The cylinder extends, when the solenoid Y is energized. In the electrical control part, as shown in Figure 3.9(a), the pushbuttons PB1 and PB2 are connected in parallel to realize the OR logic function. As shown in Figure 3.9(b) and (c), the solenoid Y gets energized through the relay, when PB1 or PB2 is pressed.

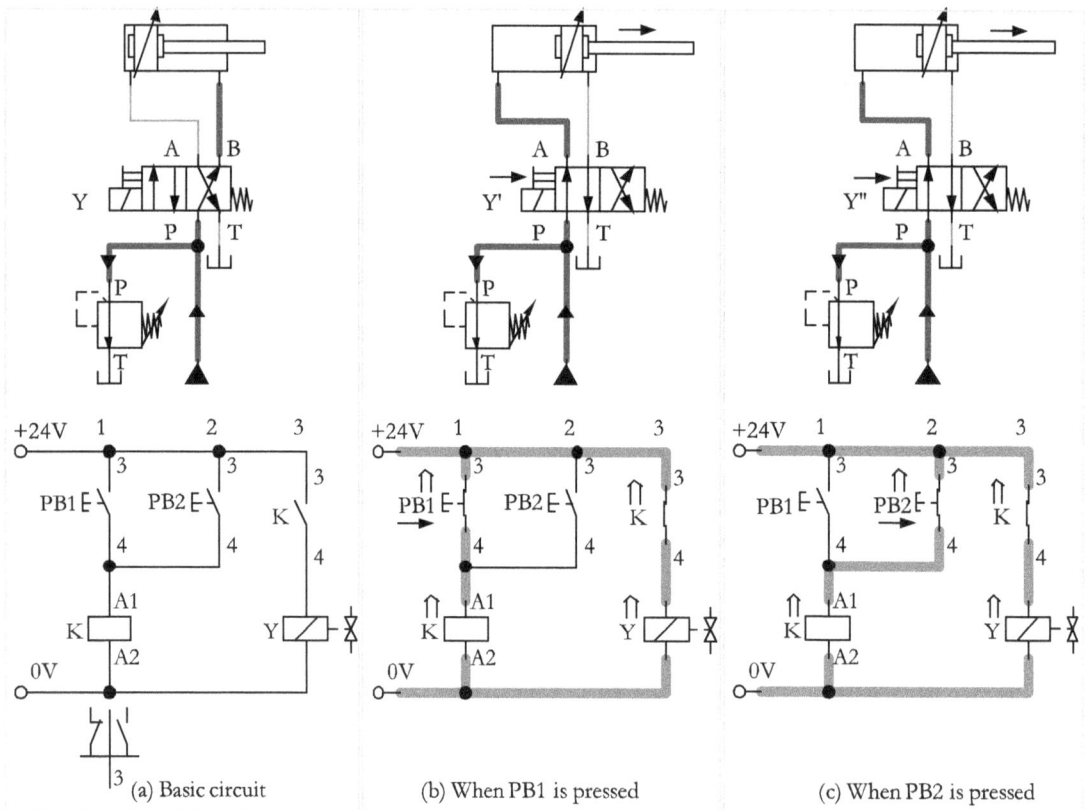

Figure 3.9 | Control of a double-acting cylinder from two locations (Example 3.4)

Example 3.5 | Two-hand safety

A large-volume double-acting hydraulic cylinder is used in a machine to punch work-pieces. A design objective is to have two-hand safety by using two pushbuttons PB1 and PB2 so installed at a distance between them far enough so that an operator cannot press both pushbuttons with one hand. That is, the operator's both hands must be engaged to operate the pushbuttons simultaneously. The cylinder should extend and carry out the punching operation when both pushbuttons are pressed simultaneously. The cylinder should retract when anyone or both pushbuttons are released. Develop an electro-hydraulic circuit to implement the control task.

Solution

The electro-hydraulic circuit in four critical positions is given in Figure 3.10. In the hydraulic part of the circuit, the double-acting cylinder is shown as controlled by a 4/2-way single solenoid spring-return valve. The cylinder extends, when the solenoid Y is energized. In the electrical control part, the pushbuttons PB1 and PB2 are connected in series to realize the AND logic function. As shown in Figure 3.10(d), the solenoid Y gets energized through the relay, when both PB1 and PB2 are pressed. For all other conditions, as shown in Figure 3.10(a), (b) and (c), the solenoid remains in the de-energized state.

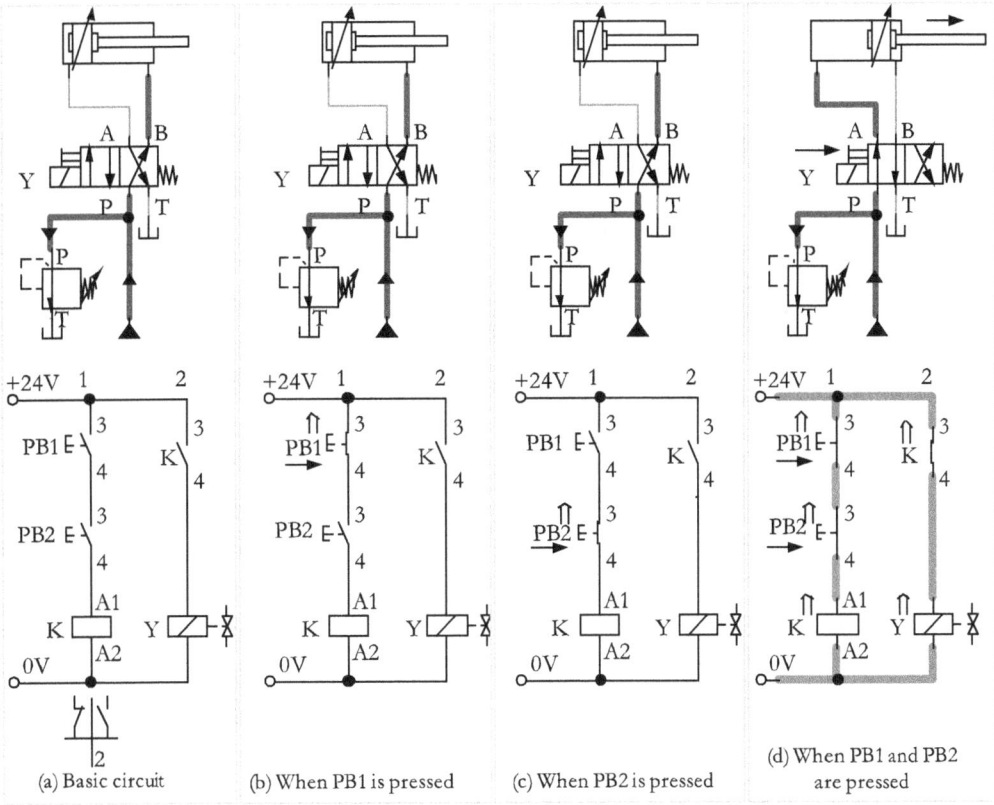

Figure 3.10 | Control of the double-acting cylinder for two-hand safety (Example 3.5)

Memory Function

A circuit with a memory function 'remembers' its last output state even after the input signal from the input device responsible for this output has been removed. That is; a momentary signal from the ON input device sets the output device to the ON state, and it remains in the set (ON) state until an OFF signal from an OFF input device is applied. After receiving the OFF input signal, the output device is reset, and it remains in the reset (OFF) state until an ON signal is applied again. A memory function can be implemented in electro-hydraulic circuits by using the electrical latching circuit or by using the double solenoid valve. The working of the electrical latching circuit is described in the following sections.

Latching Circuit, Electric

The electrical latching circuit, as shown in Figure 3.11, can be constructed with the following control components: (1) an NO pushbutton for the 'ON' or 'Start' control, (2) an NC pushbutton for the 'OFF' or 'Stop' control, and (3) a relay (K1) with a coil and contact sets. Two positions of the circuit are also given for a quick analysis of the circuit.

The following actions take place when the 'Start' pushbutton (PB1) is momentarily (or continuously) pressed, as shown in Figure 3.11(a):

- The relay coil K1 in branch 1 is energized, operating its contacts in branches 2 & 3.
- The first NO contact of the relay K1 (in branch 2) latches the 'Start' pushbutton (PB1).
- The relay coil K1 remains energized even when the pushbutton (PB1) is released.
- The second contact of K1 (in branch 3) switches the solenoid coil Y.
- The solenoid coil remains in the ON position until the 'Stop' pushbutton (PB2) is pressed.

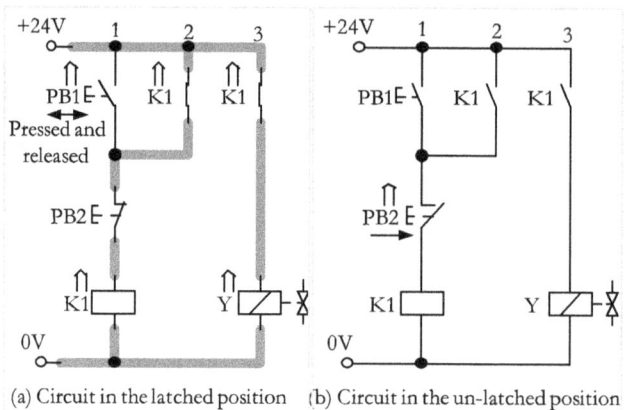

(a) Circuit in the latched position   (b) Circuit in the un-latched position

Figure 3.11 | Two positions of a latching circuit, electrical

The following actions take place when the 'Stop' pushbutton (PB2) is momentarily (or continuously) pressed, as shown in Figure 3.11(b):

- The relay coil K1 is de-energized as the relay coil circuit is interrupted.
- Each of the contacts of the relay K1 (in branches 2 & 3) returns to its normal position.
- The solenoid coil is de-energized and remains in the OFF position until the 'Start' pushbutton (PB1) is pressed again.

Example 3.6 | Latching circuit
A double-acting cylinder is to be controlled using a 4/2-way single solenoid valve. When pushbutton PB1 is pressed, the cylinder is to extend and to remain in the extended position even when PB1 is released. The cylinder is to retract to the home position when pushbutton PB2 is pressed. The cylinder is to remain in the home position even when PB2 is released. Develop an electro-hydraulic control circuit with an electrical latching circuit.

Solution
Two critical positions of the hydraulic and electrical part of the electro-hydraulic circuit when the coil Y is energized and de-energized are given in Figure 3.12. In the hydraulic part, a double-acting cylinder is controlled by a 4/2-way single solenoid valve. The valve does not exhibit memory characteristic, as it has a reset spring. The memory function can be implemented through the electrical latching circuit. The latching circuit, as shown in Figure 3.12(a), consists of a relay (K) and a NO type 'Start' pushbutton (PB1) and an NC type 'Stop' pushbutton (PB2).

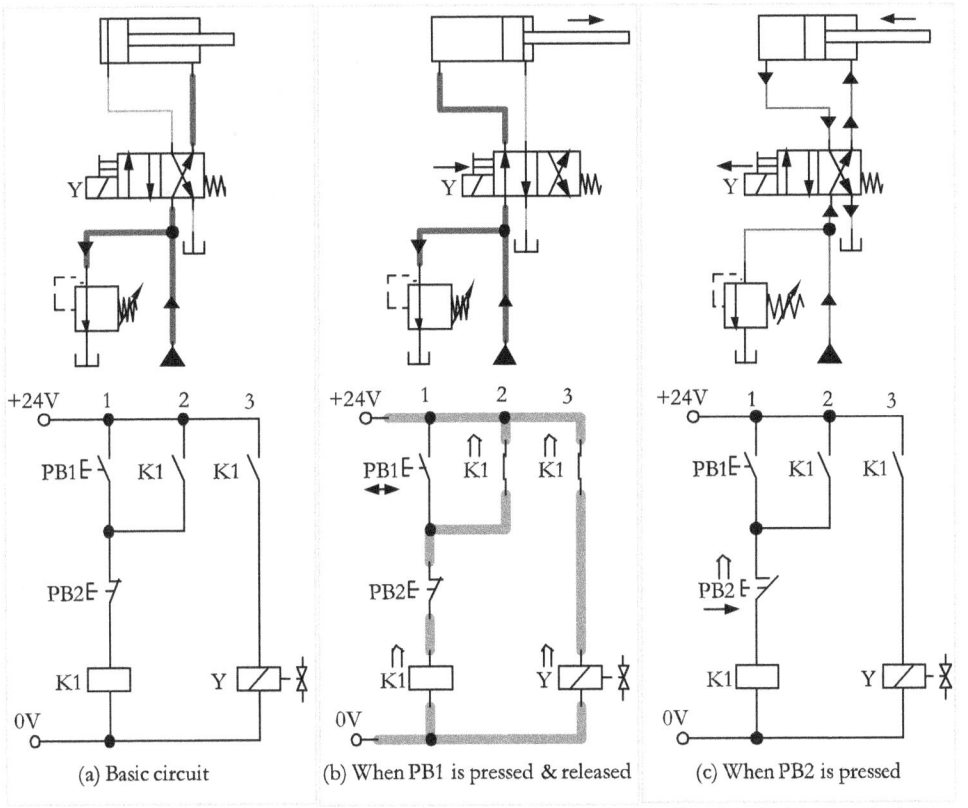

Figure 3.12 | Three critical positions of the electro-hydraulic circuit (Example 3.6)

When the 'Start' pushbutton is momentarily or continuously pressed, as shown in Figure 3.12(b), the relay coil K1 in branch 1 is energized, operating all of its contacts. The NO contact of K1, used in branch 2, latches the 'Start' pushbutton (PB1) by providing a parallel path for the current flow so that relay coil K1 remains energized even when the pushbutton (PB1) is released. This latched position of the circuit is shown in Figure 3.12(b). The contact of K1, used in branch 3, switches the solenoid valve. The cylinder extends and remains in that position until the 'Stop' pushbutton (PB2) is pressed.

When the 'Stop' pushbutton (PB2) is momentarily or continuously pressed the relay coil K1 is de-energized as the coil circuit is interrupted. All the control contacts of relay K1 return to the normal position. The opening of the relay contacts unlatches the circuit and de-energizes the solenoid. The cylinder retracts and remains in that position until the 'Start' pushbutton (PB1) is pressed again. The un-latched position of the circuit is shown in Figure 3.12(c).

Example 3.7 | Control of a hydraulic cylinder using a double solenoid valve

A double-acting hydraulic cylinder is to be controlled, using a 4/2-DC double-solenoid valve. When a pushbutton (PB1) is pressed, the cylinder is to extend and should remain in that position even when PB1 is released. When another pushbutton (PB2) is pressed, the cylinder is to retract to its home position and should remain in that position even when PB2 is released. Develop an electro-hydraulic circuit with a fixed-displacement pump for implementing the control scheme.

Solution

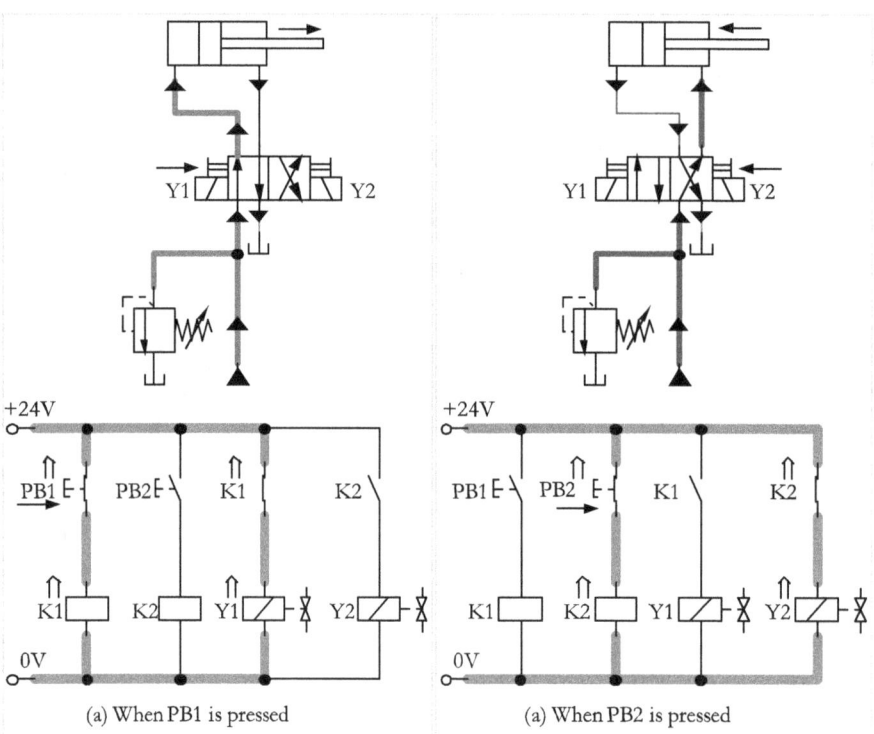

Figure 3.13 | Two positions of an electro-hydraulic circuit for the control of a cylinder using a double solenoid valve. (Example 3.7)

Apart from using an electrical latching circuit, the 4/2-way double solenoid valve can be used for implementing the memory function in the electro-hydraulic system with the fixed-displacement pump as demonstrated with the help of Figure 3.13. Let the solenoids be Y1 and Y2. Apart from this, two relays K1 and K2 are used in the circuit.

When the pushbutton PB1 is pressed, the coil Y1 is energized through the relay K1, and the solenoid valve switches over, as shown in Figure 3.13(a). The cylinder extends and remains in the forward end position until a control signal is applied in the opposite direction. When the pushbutton PB2 is pressed, the coil Y2 is energized through the relay K2, and the solenoid valve switches over, as shown in Figure 3.13(b). The cylinder retracts to its home position until a signal is applied to the coil Y1 again.

Sensors
Sensors have taken the place of human senses in today's automated industrial operations. They can detect the presence of objects. Next, they can measure and process various changes that occur at the production sites. The sensor can work either by the actual physical contact with an object or by the movement of the object in its proximity. Accordingly, the sensors are classified as contact-type sensors (e.g. limit switch) and contactless-type sensors (e.g. proximity sensors).

The contact-less type sensor uses the magnetic or electrostatic or optical medium to realize the sensing function. Any disturbance of the physical medium produces a signal at the output of the sensor. That is; the contactless sensor can generate a measurable output signal in response to the changes in its physical condition.

It may be noted that a variety of sensors is devised to meet the varied demands of industrial production systems.

Limit Switch
A limit switch is a contact-type sensor comprising a set of switching contacts (NO/NC/CO type), a roller-operated plunger and return springs as shown in the diagram of Figure 3.14. The roller lever is mechanically linked to the contacts. It is usually actuated mechanically by a moving element, such as a cylinder piston, in the associated machine to indicate a particular position of the moving element. The electrical contact is established or interrupted using the actuating force acting on the roller lever.

The limit switch produces an electrical signal upon detecting the position of the mechanical member to be detected.

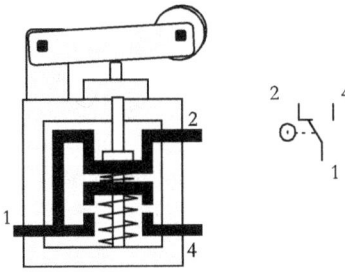

Figure 3.14 | A cross-sectional view of a limit switch with a changeover (CO) contact

Limit Switches are particularly suited for applications that require mechanical strength or environmental resistance. However, the service life of a limit switch tends to be low as the switch detects the presence of an object by the physical contact.

Reed Switch

A reed switch is also known as a magnetically-actuated proximity switch. It consists of two metal strips (reeds) acting as switching contacts. It is hermetically sealed in a glass tube filled with an inert gas. This encasing is done to prevent the corrosion of its contacts. This unit is further encapsulated in epoxy resin casing. This covering of the reeds prevents its possible mechanical damage. The reed switch is usually provided with a light-emitting diode (LED) to indicate its switching status. It is designed for mounting on a cylinder. It reacts to the magnetic fields of the permanent magnets provided on the cylinder piston. That is; the reed switch closes when the piston is adequately near for the magnetic field to actuate its contacts. The schematic diagram of Figure 3.15 shows the reed switches positioned at the cap-end position and head-end position of the cylinder. For the retracted position of the cylinder, the reed switch at the cap-end is closed, and that at the head-end is open, as illustrated in Figure 3.15.

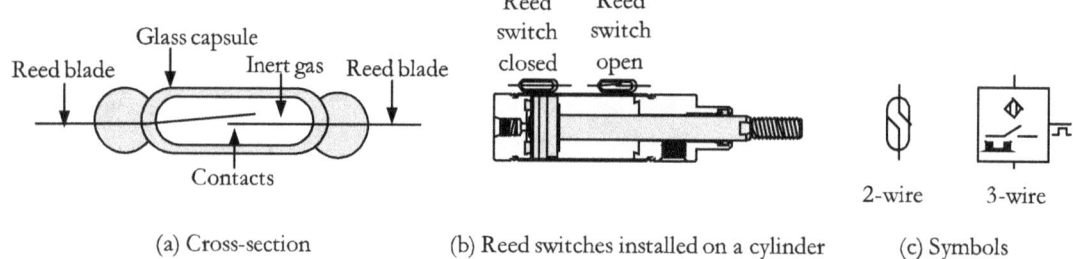

(a) Cross-section   (b) Reed switches installed on a cylinder   (c) Symbols

Figure 3.15 | A schematic diagram showing two reed switches mounted on a hydraulic cylinder

The primary reed switch consists of only two wires – one for connecting to the positive (+) lead of the power supply and the other for taking out the signal output. The three-wire reed switch with an LED indicator consists of three wires. The first one is for connecting to the positive (+) lead of the power supply, the second one is for taking out the signal output, and the third one is for connecting to the (-) lead. The connection to the negative terminal is necessary for the LED indication.

Proximity Sensors

A proximity sensor, as shown in Figure 3.16, is a contactless-type sensor that detects the presence of an object using a detection system and converts this information into a corresponding electrical signal. The proximity sensors use static switches without moving contacts, and hence, they provide a high-speed response. One type of detection system uses the eddy currents that are generated in a metallic sensing object by the interaction of the detection system and the object. Another type detects the changes in the electrical capacity of the capacitor in the detection system when approaching an object. Yet another type detects objects through a variety of optical properties. Accordingly, there are three basic types of proximity sensors. They are: (1) Inductive-type sensors, (2) Capacitive-type sensors, and (3) Optical-type sensors. The following sections briefly explain these types of sensors.

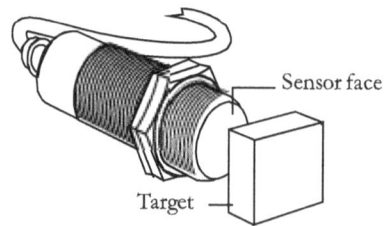

Figure 3.16 | A schematic diagram of a proximity sensor and a target

The most important factor for a proximity sensor is its maximum sensing distance. It is the maximum distance at which the switching action of the sensor can take place. Proximity sensors can be used for applications where the temperature varies in a wide range typically from -40 ⁰C to 200 ⁰C. Proximity sensors are ideal for use in dirty or wet environments.

As the proximity sensors offer superior performance, they are replacing the traditional limit switches. Their utilization in an existing piece of equipment helps to upgrade the speed and reliability of the equipment. Some of the applications of the proximity sensors are found in systems for inspecting high-speed table movements, positioning a welding site, performing the level control of non-conductive liquids, and detecting aluminum components and bottle caps.

Inductive Proximity Sensor
Inductive proximity sensors are widely used in modern high-speed industrial and process control systems for the detection of metal objects. An inductive proximity sensor consists of the following blocks: (1) an oscillator circuit, (2) a switching circuit, (3) an amplifier, and (4) an output stage, all housed in a resin-encapsulated body. Figure 3.17 shows the block diagram of the inductive proximity sensor.

A part of the oscillator circuit is a coil capable of producing high-frequency magnetic oscillations in the active switching area in front of the sensing face of the sensor when the rated voltage is applied to the sensor. If any metallic object is brought near to the active switching area without necessarily making physical contact, eddy currents are generated in the object. The eddy currents are converted into heat. The generated heat represents a loss of energy. Note that this loss draws energy from the oscillator. As a result, the oscillations are weakened. The switching circuit converts this state of the oscillator into a clear signal through the switching stage. Finally, the output signal is amplified and delivered to the load circuit. The sensing range of inductive proximity sensors is usually small, typically up to 12 mm.

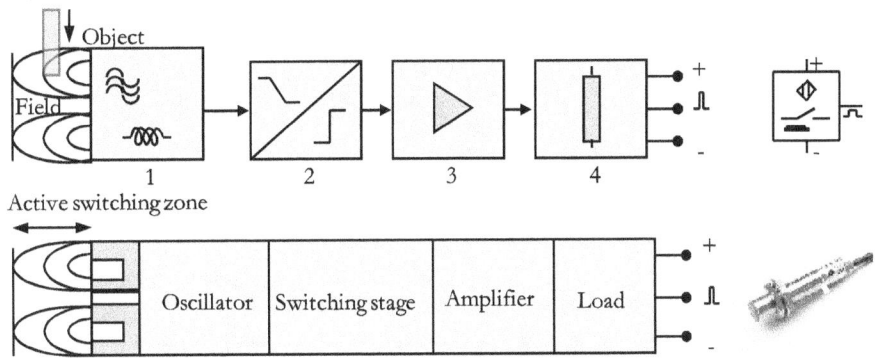

Block diagrams of inductive proximity sensor
Figure 3.17 | A block diagram of an inductive proximity sensor

Capacitive Proximity Sensor
Capacitive proximity sensors are often used in applications that cannot be solved with other types of sensors. A capacitive proximity sensor consists of the following blocks: (1) an oscillator, (2) a switching circuit, (3) an amplifier, and (4) an output stage. Figure 3.18 shows the block diagram of the capacitive proximity sensor.

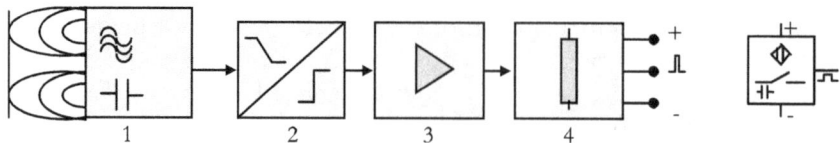

Figure 3.18 | A block diagram of the capacitive proximity sensor

The oscillator in the capacitive proximity sensor comprises a capacitor with two electrodes. One electrode is the sensing surface of the sensor, and the second one is the sensing object itself. An electrostatic field is created in the active switching area of the sensor when the rated voltage is applied to the sensor. If the object is brought into the active switching area, the capacitance of the oscillator circuit gets altered. The change in the capacitance of the oscillator circuit triggers the switching circuit to produce an output signal. The capacitive sensor can respond to the change in the strength of the dielectric medium surrounding the active face. Finally, the output signal is amplified and delivered to the load circuit. Remember, capacitive sensors can sense almost any object. The sensing range of capacitive proximity sensors is typically up to 25 mm.

Optical Proximity Sensors

Optical (or photoelectric) sensors offer the non-contact-type sensing of almost any object with a sensing range typically up to 10 meters. An optical proximity sensor employs optical means for sensing objects. This type of sensor consists of a transmitter (or emitter) and a receiver. Usually, the transmitter is a Light Emitting Diode (LED), and the receiver is a phototransistor. The transmitter emits straight moving infrared rays when an electrical current is passed through the LED, and the receiver reacts to the infrared rays. Optical sensors are available as diffuse reflective, through-beam, and retro-reflective models. The following sections explain two of these types of optical sensors.

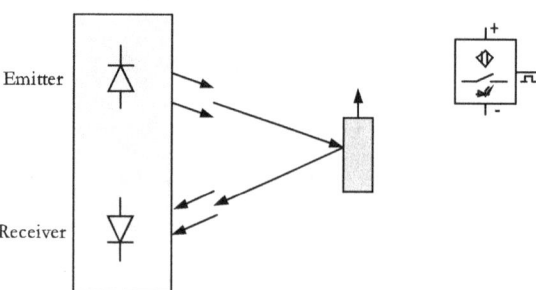

Figure 3.19 | A block diagram showing the action of a diffuse reflective optical proximity sensor

In the diffuse reflective type of the optical sensors, the transmitter and receiver are arranged in the same housing as shown in the schematic diagram of Figure 3.19. Here, the object diffusively reflects a percentage of the emitted rays when interrupted by an object, thereby activating the receiver. The receiver detects the interruption of the light medium and converts it to an electrical output. This type of photoelectric sensors is preferable for the general-purpose applications, particularly where the detected object is accessible only from one direction.

In the through-beam optical proximity sensor, the transmitter and the receiver are arranged in line with the emitted infrared rays from the transmitter to hit the receiver, as shown in Figure 3.20. An object passing between the transmitter and the receiver can interrupt the infra-red rays, causing the receiver to generate an output signal.

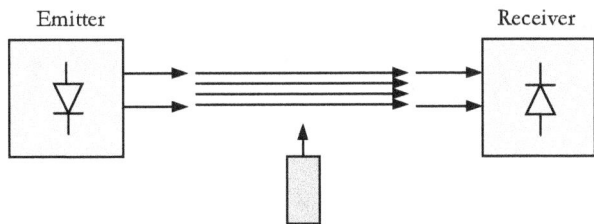

Figure 3.20 | A block diagram showing the action of a through-beam optical proximity sensor

Example 3.8 | Semi-automatic operation of a double-acting cylinder using 4/2-DC single-solenoid valve and limit switch

A double-acting cylinder is to extend when a pushbutton is pressed. On reaching the end position, the cylinder is to retract automatically. A 4/2-DC single-solenoid valve is used as the final control element. Develop an electro-hydraulic control circuit to implement the control task using a limit switch for the semi-automatic operation of the cylinder.

Solution

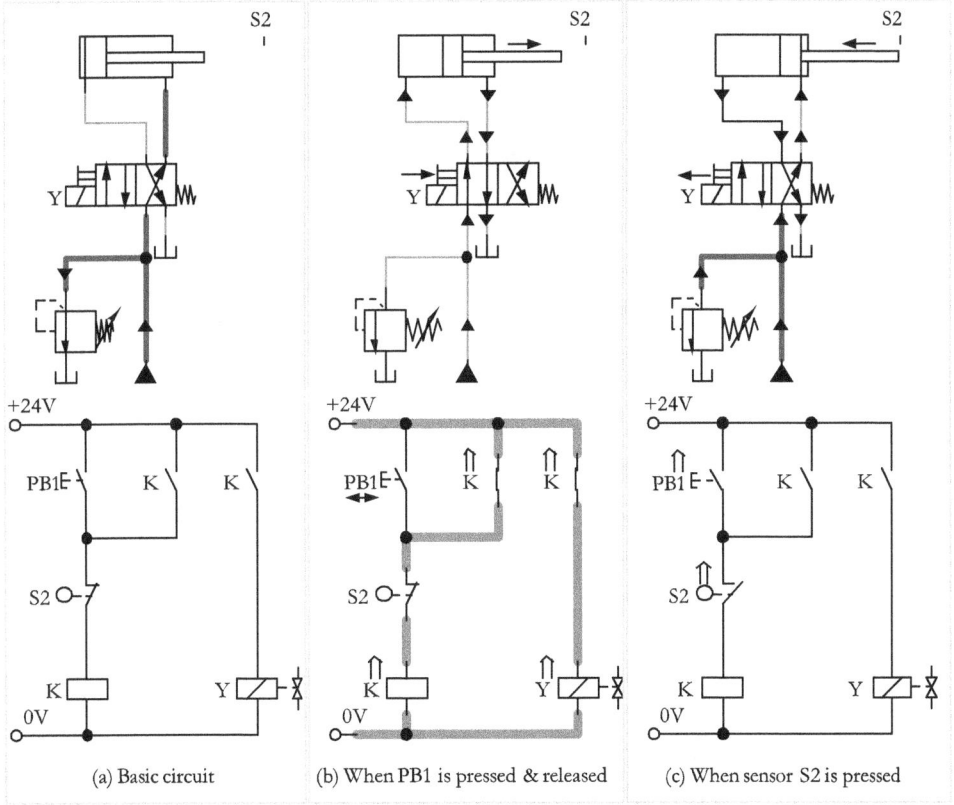

Figure 3.21 | Auto-return of the double-acting cylinder using a limit switch (Example 3.8)

The double-acting cylinder is controlled by a 4/2-DC single-solenoid, spring-return valve [Figure 3.21(a)]. The electrical circuit is latched when pushbutton PB1 is actuated. The valve 1.1 remains in the actuated position even when the pushbutton PB1 is released [Figure 3.21(b)]. The cylinder then starts moving forward. When the fully extended position of the cylinder is reached, it automatically actuates limit switch S2. Actuation of the limit switch causes interruption of the electrical circuit [Figure 3.21(c)]. The cylinder then starts retracting automatically.

Example 3.9 | Semi-automatic operation of a double-acting cylinder using a 4/2-DC single-solenoid valve and proximity sensor

A double-acting cylinder is to extend when a pushbutton is pressed. On reaching the end position, the cylinder is to retract automatically. A 4/2-DC single-solenoid valve is used as the final control element. Develop an electro-hydraulic control circuit to implement the control task using a proximity sensor for the semi-automatic operation of the cylinder.

Solution

The double-acting cylinder is controlled by a 4/2-DC single-solenoid, spring-return valve [Figure 3.22(a)]. The position of the circuit when pushbutton PB1 is pressed and then released is given in Figure 3.22(b). The cylinder extends to its forward-end position and influences the proximity sensor S2 automatically. Relay K2 is connected across the proximity sensor to convert voltage outputs of the proximity sensor to corresponding contact operations. An NC contact of K2 is used to interrupt the latching circuit when the proximity sensor is sensing. This position of the circuit is shown in Figure 3.22 (c). The cylinder then starts retracting automatically.

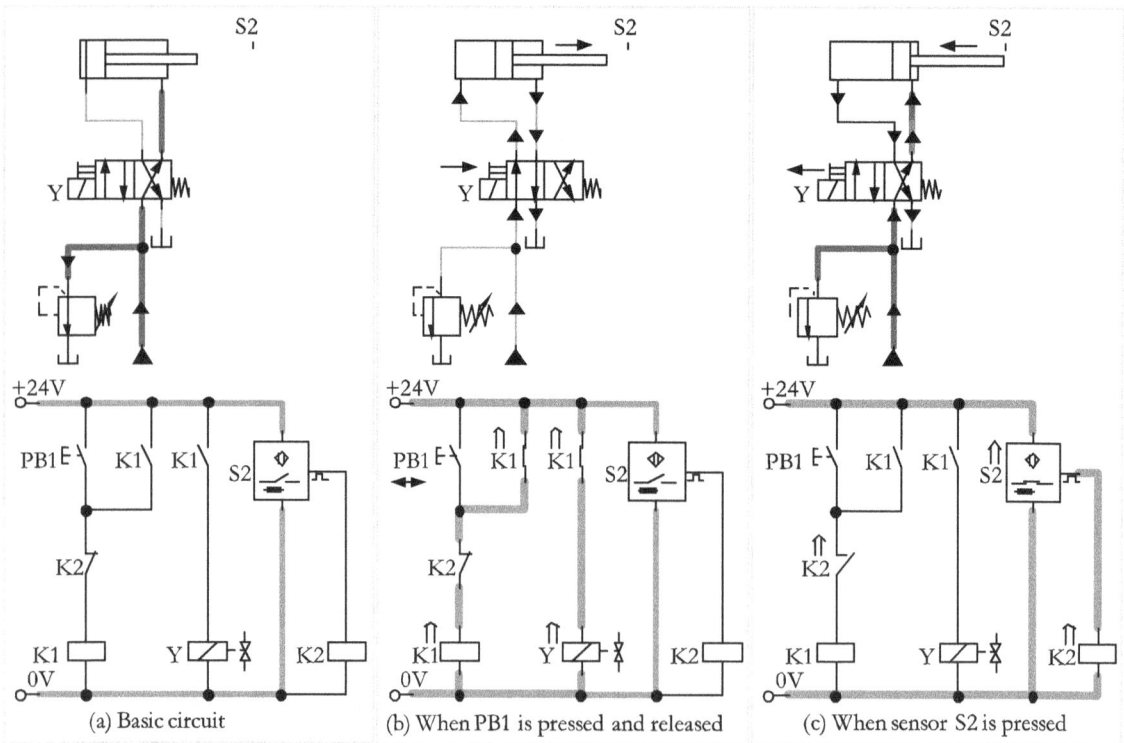

Figure 3.22 | Auto-return of the double-acting cylinder using a proximity sensor (Example 3.9)

Example 3.10 | Semi-automatic operation of a double-acting cylinder using 4/2-DC double-solenoid valve and limit switch

A double-acting cylinder is to extend when a pushbutton is pressed. On reaching the extreme end position, the cylinder is to retract automatically. A 4/2-DC double-solenoid valve is used as the final control element. Develop an electro-hydraulic control circuit to implement the control task using a limit switch for the semi-automatic operation of the cylinder.

Solution

The double-acting cylinder is controlled by the 4/2-DC double-solenoid valve with solenoids Y1 and Y2 [Figure 3.23(a)].

The solenoid Y1 is energized when the pushbutton PB1 is pressed momentarily (or continuously), as shown in Figure 3.23(b). The valve is actuated to its left envelope and remains in that position even when PB1 is released. The cylinder then extends.

When the fully extended position of the cylinder is reached, it automatically actuates limit switch S2, as shown in Figure 3.23(c). The valve is actuated to its right envelope and remains in that position even when the limit switch S2 is released. The cylinder then retracts automatically.

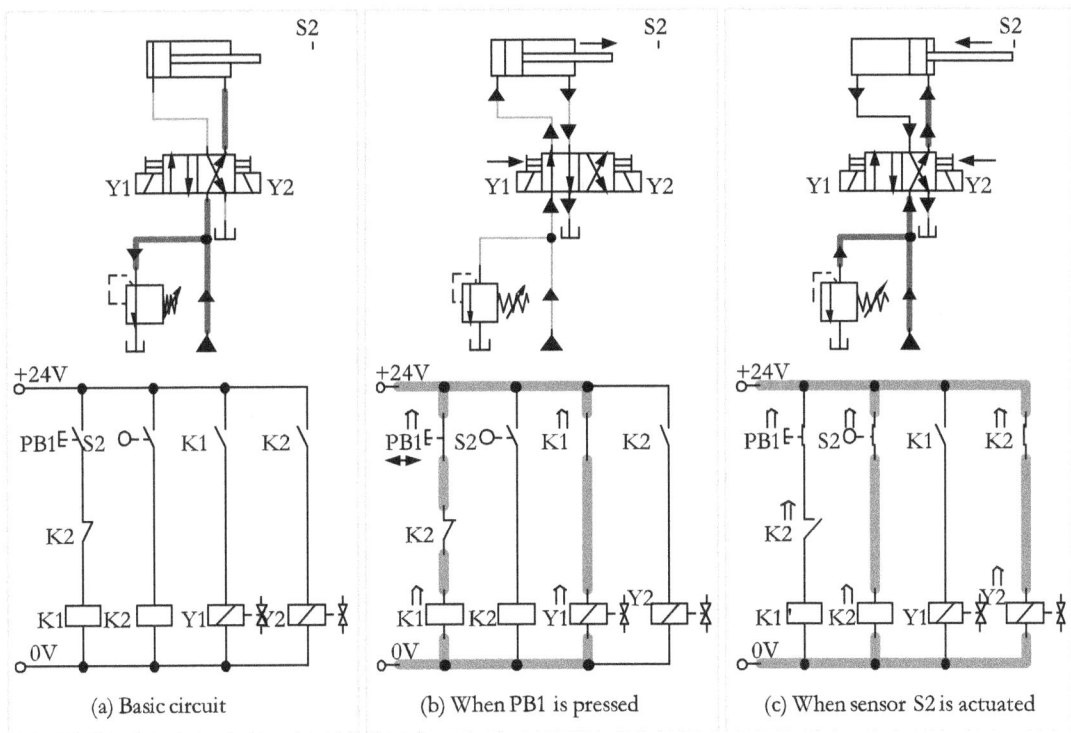

Figure 3.23 | Auto-return of the double-acting cylinder using a limit switch (Example 3.10)

Example 3.11 | Semi-automatic operation of a double-acting cylinder using a 4/2-DC double-solenoid valve and proximity sensor

A double-acting cylinder is to extend when a pushbutton is pressed. On reaching the extreme end position, the cylinder is to retract automatically. A 4/2-DC double-solenoid valve is used as the final control element. Develop an electro-hydraulic control circuit to implement the control task using a proximity sensor for the semi-automatic operation of the cylinder.

Solution
The double-acting cylinder is controlled by the 4/2-DC double-solenoid valve with solenoids Y1 and Y2 [Figure 3.24(a).

The solenoid Y1 is energized when the pushbutton PB1 is pressed momentarily, as shown in Figure 3.24(b). The valve is actuated to its left envelope and remains in that position even when PB1 is released. The cylinder then extends.

When the fully extended position of the cylinder is reached, it automatically activates the proximity sensor S2, as shown in Figure 3.24(c). The valve is actuated to its right envelope and remains in that position even when the sensor S2 is released. The cylinder then retracts automatically.

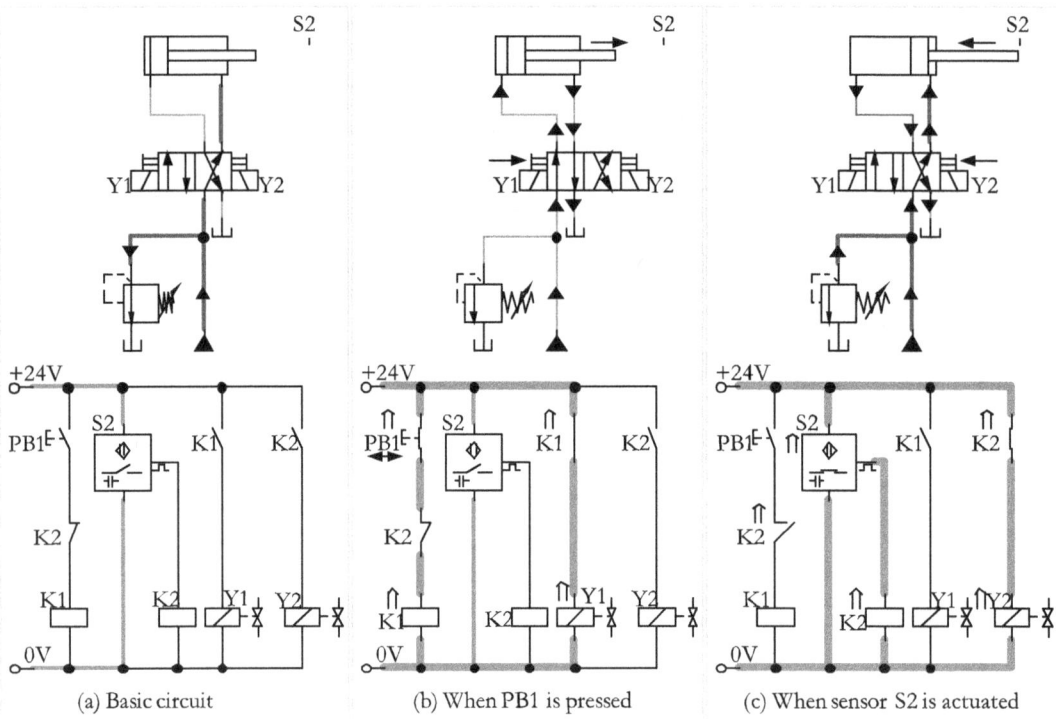

Figure 3.24 | Auto-return of the double-acting cylinder using a proximity sensor (Example 3.11)

Time-delay Relays (Timers)

Time-delay relays (or timers) are control devices, used to obtain a specified time delay between the work operations in industrial systems. For example, a timer can be used for getting the specified duration of the time delay between the forward and return strokes of a hydraulic cylinder or between the motions of two actuators. The timer is a control device that generates a delayed output signal in a circuit when the input signal is applied to it.

Electronic timers are well-known in the present-day industrial control systems. An electronic timer mainly consists of a coil, control contacts, and an electronic circuit. The delay time can be set on the timer using the potentiometer in the timer. The contact operation can be delayed through the electronic circuit when the coil is energized or de-energized. Accordingly, there are two basic types of timers. They are: (1) On-delay timer and (2) Off-delay timer. The following sections explain the operations of these types of timers.

On-delay Timer: An on-delay timer consists of a coil, contacts, and an electronic delay circuit. This type of timer delays the operation of its contacts for, say 't' seconds, when the coil is energized (ON), but the timer brings back its contact sets immediately to their normal positions when the coil is de-energized (OFF). Figure 3.25 gives the symbol and timing diagram of the on-delay timer.

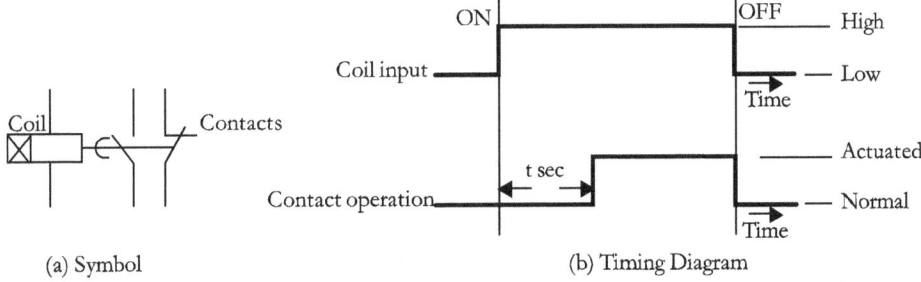

Figure 3.25 | A symbolic representation of an on-delay timer and its timing diagram

Off-delay Timer: An off-delay timer consists of a coil, control contacts, and an electronic delay circuit. It operates its contacts immediately when the coil is energized (ON). Now, when the coil is de-energized, the timer maintains the state of its contacts for the specified duration of time, say 't' sec. That is; it delays the return motion of its contact sets when the coil is de-energized. Figure 3.26 gives the symbol and timing diagram of the off-delay timer.

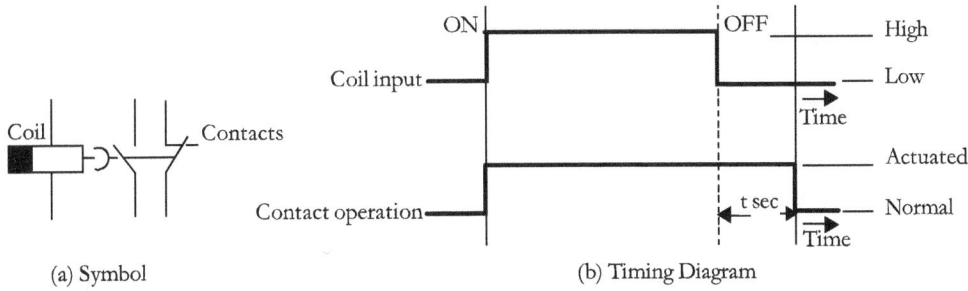

Figure 3.26 | A symbolic representation of an off-delay timer and its timing diagram

Example 3.12 | Control of a double-acting cylinder using a timer

A double-acting hydraulic cylinder is to extend when a pushbutton is pressed momentarily. It is to remain in the extended position for a period (say t seconds), and then to return automatically. The final forward end position of the cylinder is detected by a limit switch. A 4/2-DC single-solenoid valve is used as the final control element. Develop an electro-hydraulic circuit with a fixed-displacement pump to implement the control function.

Solution

The double-acting hydraulic cylinder is powered by the fixed-displacement pump and controlled by the 4/2-DC single-solenoid, spring return valve [Figure 3.27(a)]. The electrical circuit is latched through a relay (K1) when the pushbutton (PB1) is actuated. The cylinder then extends. When the fully extended position of the cylinder is reached, it automatically actuates the limit switch (S2). Figure 3.27(b) shows this position of the circuit.

An on-delay timer (T) is used to obtain the delayed return motion of the cylinder. The required delay time ('t' sec) can be set on the timer. The limit switch S2 controls the electric supply to the timer coil (T). That is, as soon as the cylinder actuates the limit switch S2, the timer coil is energized. The latching circuit is interrupted by the timer contact after the elapse of the set delay time. This interruption of the circuit causes the return movement of the cylinder, as shown in Figure 3.27(c).

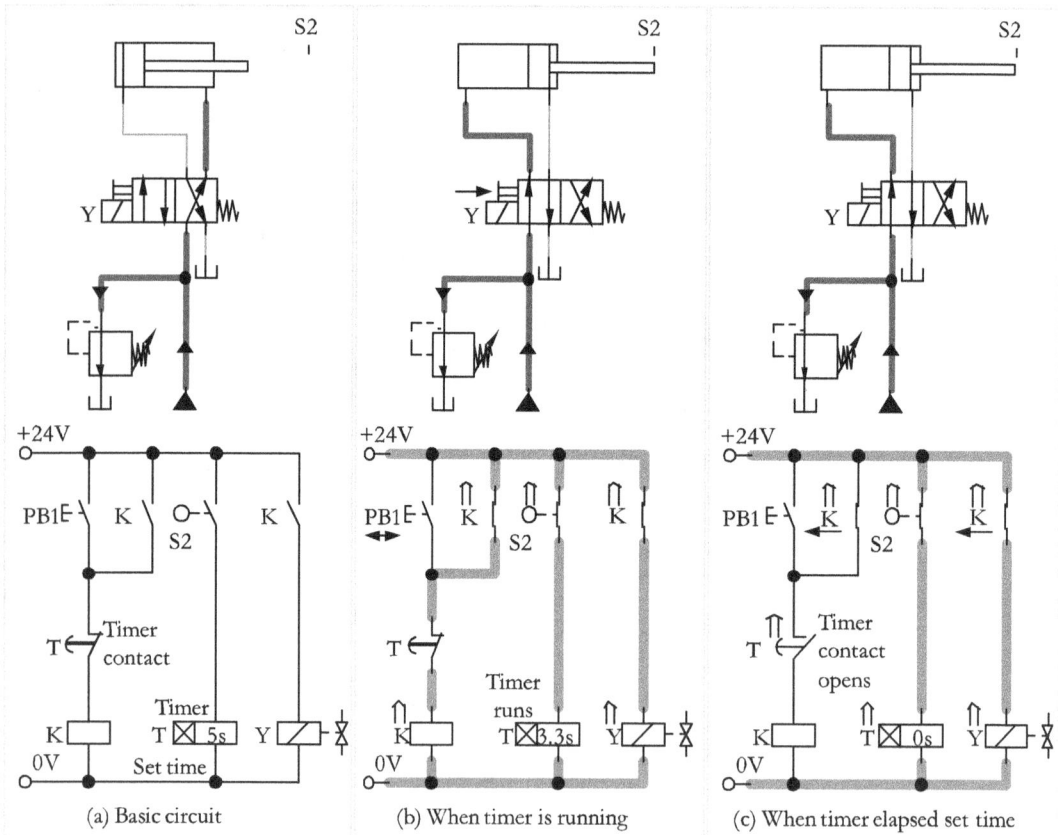

Figure 3.27 | Critical views of an electro-hydraulic circuit for the control of a double-acting cylinder using a timer. (Example 3.12)

Example 3.13 | Control of a double-acting cylinder using a timer

A double-acting hydraulic cylinder is to extend when a pushbutton is pressed momentarily. It is to remain in the extended position for a period (say t seconds), and then to return automatically. The final forward end position of the cylinder is detected by a proximity sensor. A 4/2-DC single-solenoid valve is used as the final control element. Develop an electro-hydraulic circuit with a fixed-displacement pump to implement the control function.

Solution

A latching circuit is used to obtain the necessary memory function. The position of the circuit when pushbutton PB1 is pressed and then released is given in Figure 3.28(a).

The cylinder extends to its forward-end position and actuates limit switch S2 automatically, as shown in Figure 3.28(b).

As the return motion is to be delayed, an on-delay timer is used to obtain the necessary time delay. The required time delay can be set on the timer. Limit switch S2 controls the timer coil T. After the set delay, the timer contact interrupts the latching circuit, thus causing the return motion of the cylinder, as shown in Figure 3.28(c).

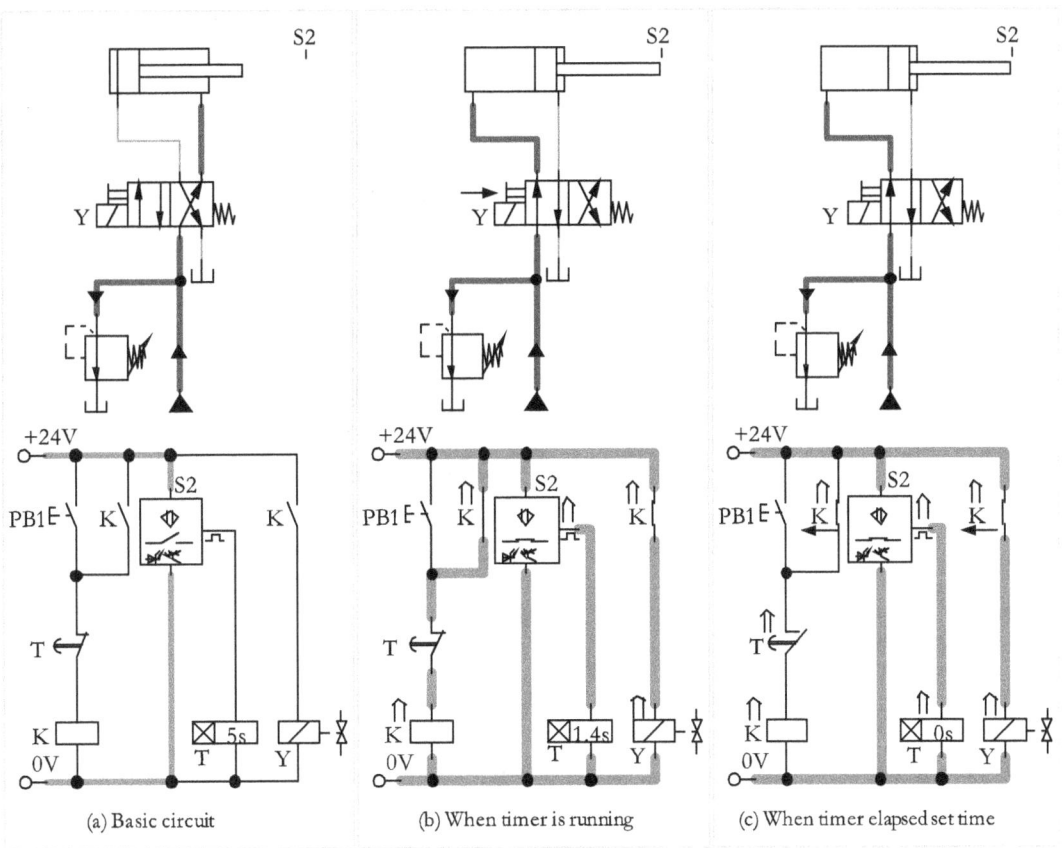

Figure 3.28 | The control of a double-acting cylinder using a timer (Example 3.13)

Two-hand Safety Operation with Anti-tie-down Feature

A machine for pressing, cutting or other similar operations is usually designed with a two-hand safety feature using two pushbuttons (PBs) installed in such a way that both of them cannot be pressed simultaneously by one hand. This arrangement means that an operator can control the machine only by pressing the two pushbuttons simultaneously. Moreover, he has to use both his hands to do this for safety purpose. This arrangement ensures that both of his hands are engaged in pressing the pushbuttons, and thus his hands cannot be on the machine while it is operating.

Further, the machine should not operate when one of the two pushbuttons is tied down permanently by some means. On many occasions, machine operators have done this to free one of his hands to adjust the work-piece into the machine while it is operating. This way of working is a hazardous practice. Anti-tie down and anti-repeat circuits can be used to ensure that both switches must be OFF and must be pressed simultaneously within a short duration of time usually within half a second to operate the machine just for one cycle. For the next cycle, these pushbuttons must be rereleased and then pressed again.

Pressure Switch

A pressure switch is a hydraulic–electric (P/E) signal converter. It is a control element that automatically senses a change in the applied pressure through a pressure sensing element and operates an electrical switching element when a predetermined pressure point is reached. The pressure-sensing element is that part of a pressure switch that moves due to the change in pressure. There are three types of sensing elements commonly used in pressure switches. They are: (1) Diaphragm type, (2) Bourdon tube type, and (3) Sealed piston type.

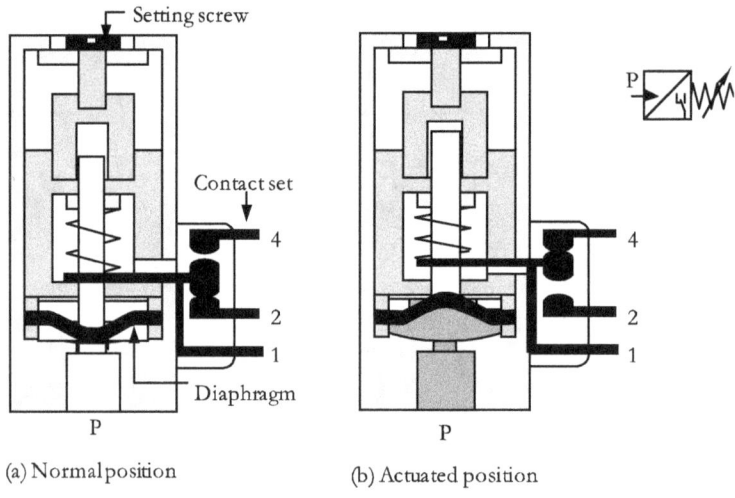

Figure 3.29 | Cross-sectional view of a pressure switch

The diaphragm type, as shown in Figure 3.29, consists of a metal diaphragm. When the preset pressure is reached at the inlet port of the pressure switch, the diaphragm expands and pushes the spring-loaded plunger. This force, in turn, operates the contacts. It can typically operate from vacuum pressure up to a pressure of 10 bar [145 psi].

The Bourdon tube type employs a hollow tube in a semi-circular shape whose design is such that an increase in pressure tends to straighten it. The resulting force is sufficient to actuate an integrated snap-action switch. It can typically operate from 3.5 bar [50 psi] to 1200 bar [18000 psi].

The sealed piston-type consists of an O-ring-type-sealed piston that acts on a snap action switch. It can typically operate from 1 bar [15 psi] to 800 bar [12000 psi].

Example 3.14 | Stamping device
Components are to be stamped using a stamping device. A double-acting cylinder pushes the die attached down to a fixture when a pushbutton (PB1) is pressed. The die is to return to the initial position upon reaching sufficient stamping pressure is sensed by a pressure switch. Develop an electro-hydraulic control circuit to implement the control task for the stamping operation.

Solution

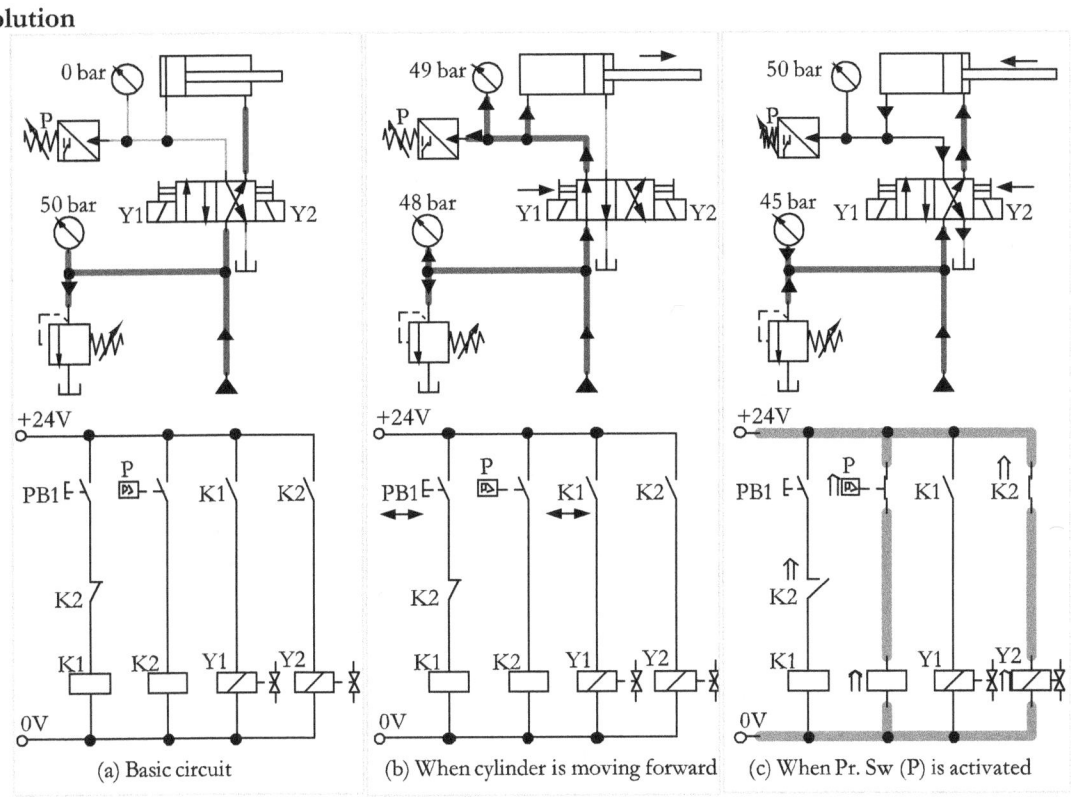

Figure 3.30 | Electro-hydraulic circuit for the stamping device (Example 3.14)

The cylinder is controlled by a 4/2 DC valve, as shown in Figure 3.30(a). A pressure switch is connected to the line leading to the piston side of the cylinder. When the pushbutton PB is pressed, the solenoid coil Y1 is energized through the relay K1 [Figure 3.30(b)]. The DC valve switches over, and the cylinder extends. When the preset switching pressure is reached in the supply line of the cylinder piston side, pressure switch P is activated [Figure 3.30(c)]. Consequently, relay coil K2 and solenoid coil Y2 are energized. The DC valve switches and then the cylinder retracts.

Example 3.15 | Cyclic operation of a double-acting cylinder

When a 'Start' pushbutton is pressed, a double-acting cylinder is to perform a continuous back-and-forth motion until a 'Stop' pushbutton is pressed. The cylinder should stop in the retracted position always. A 4/2-DC double solenoid valve is used as the final control element, and reed sensors are used for position sensing. Develop an electro-hydraulic control circuit for implementing the fully automatic operation of the cylinder.

Solution

The electro-hydraulic circuit for the cyclic operation of a double-acting cylinder controlled by a 4/2-double-solenoid valve is given in Figure 3.31(a). Reed switches S1 and S2 are positioned for actuation by the cylinder at the retracted and extended positions, respectively. The reed switch S1 is actuated in the initial position, which is represented in the drawing with its contact in the actuated position and an arrow alongside.

The fully automatic cyclic operation of the cylinder can be obtained simply by using the output signal of the reed switch S1 controlling coil Y1 through relay coil K1, and the output signal of reed switch S2 controlling coil Y2 through relay coil K2.

The 'start' and 'stop' controls of the cyclic operation can be implemented by using a latching circuit using a relay K and controlled by pushbuttons PB1 (Start) and PB2 (Stop). A contact of the relay K is connected in the supply circuit (or in the output circuit) of the reed switch S1 to obtain the necessary 'start' and 'stop' controls. The position of the circuit when the reed switch S2 is activated is shown in Figure 3.31(b).

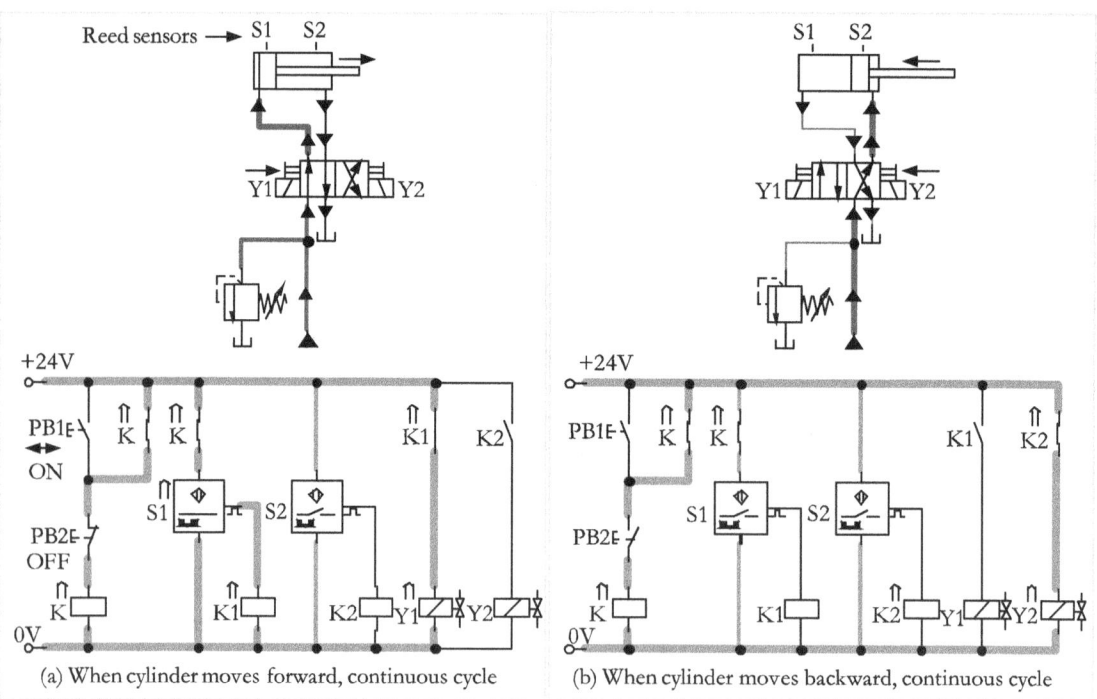

Figure 3.31 | Circuit for the cyclic operation of a double-acting cylinder (Example 3.15)

Counters

An electrically operated counter consists of a coil, associated circuits and contacts, a reset coil, manual reset, release button and a display window. Pressing the release button of the counter and entering the desired count value set the predetermining counter. The predetermined count value is displayed in the window. An up counter counts electrical signals upwards from zero. For each electrical counting pulse input to an up-counter coil, the count value is incremented by 1. When the predetermined value has been reached, the relay picks up, and the contact set is actuated. A down counter counts electrical signals downwards from a preset number. If the count value of zero is reached the relay picks up and the contact set is actuated. The counter can be reset manually by pressing the reset button or electrically by applying a reset pulse to the reset coil. The predetermined value is maintained when the counter is reset. The symbol of an up counter is given in Figure 3.32.

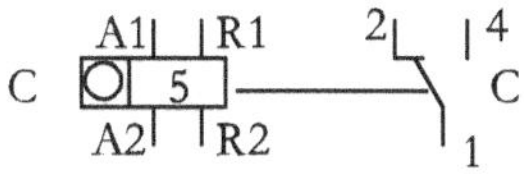

Figure 3.32 | Symbol of a counter, electric

Example 3.16 | Control circuit using a counter

When a 'Start' pushbutton is pressed, a double-acting cylinder is to perform a continuous back-and-forth motion. The cylinder should stop automatically after performing, say, 5 cycles of operation. Develop an electro-hydraulic control circuit using a down counter.

Solution

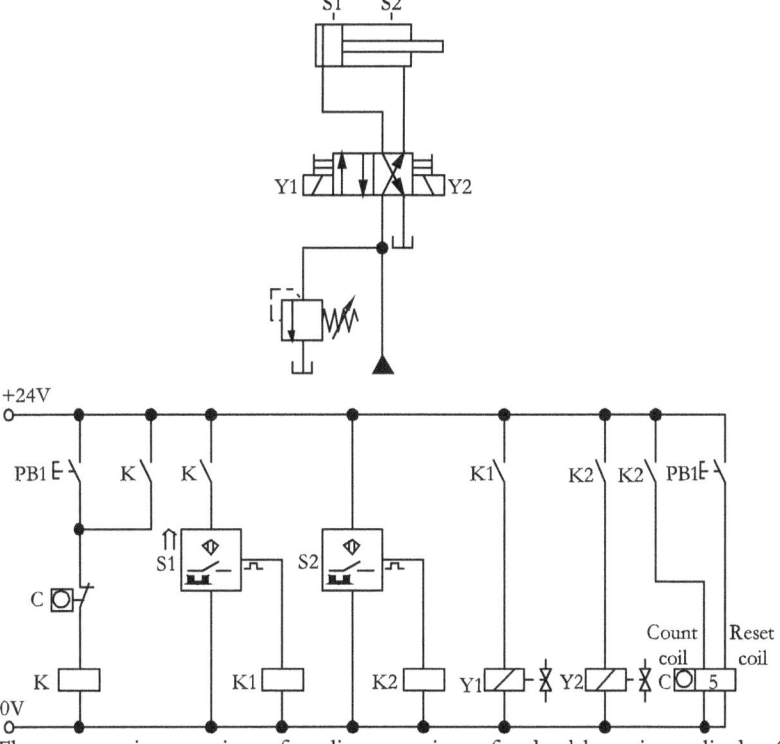

Figure 3.33 | The automatic stopping of cyclic operation of a double acting cylinder (Example 3.16)

The electro-hydraulic circuit for the cyclic operation of a double-acting cylinder controlled by a 4/2-double-solenoid valve is given in Figure 3.33. Reed switches S1 and S2 are positioned for actuation by the cylinder at the retracted and extended positions, respectively.

The forward motion of the cylinder is controlled by solenoid coil Y1 through contact of relay K1 which is in turn controlled by the contact of K and the contact of S1 in a series connection. The return motion is controlled by solenoid Y2 through a contact of K2 which is, in turn, controlled by the contact of S2. A signal pulse from sensor S2 is input to the counter coil through the relay K2, in each cycle. The NC contact of the counter is used to interrupt the latching circuit and to stop the cyclic operation after the set number of cycles of operation is complete.

Objective Type Questions

1. Mark the <u>correct</u> statement.
    a) A Normally Open (NO) contact permits energy flow in its normal position.
    b) A Normally Open (NO) contact inhibits energy flow in its actuated position.
    c) A Normally Closed (NC) contact permits energy flow in its normal position.
    d) All of the above.

2. A memory function can be implemented in an electro-hydraulic system by using a:
    a) Logic circuit
    b) Latching circuit
    c) Timer circuit
    d) Single-solenoid, spring-return valve

3. Which of the following is the <u>correct</u> statement?
    a) The limit switch is an analog device.
    b) An Inductive-type sensor can sense non-metals.
    c) A Capacitive-type sensor can sense almost all objects.
    d) The maximum sensing distance of optical sensors is very small.

4. Which of the sensors is capable of sensing only metals?
    a) Inductive-type proximity sensor
    b) Capacitive-type proximity sensor
    c) Optical-type proximity sensor
    d) None

5. Mark the <u>incorrect</u> statement.
    a) A relay is an electromagnetically operated switch.
    b) A reed switch is a magnetically actuated proximity switch.
    c) A proximity sensor generates a signal in response to the disturbance to its physical medium.
    d) An electronic timer is a sensor with an electronic oscillator circuit.

Review Questions

1. What is a solenoid valve, as used in hydraulic systems?
2. Explain how the actuating force in an electro-hydraulic valve is developed through a solenoid.
3. Write a brief note on hydraulic solenoid valves.
4. Explain the actuation mechanism in a conventional electro-hydraulic valve.

5. Describe the construction and working of a solenoid-actuated valve with a diagram.
6. Explain the operation of a 3/2-DC single-solenoid hydraulic valve.
7. Explain the operation of a 4/2-DC single-solenoid hydraulic valve.
8. Describe the functioning of a 4/2-DC double-solenoid hydraulic valve.
9. Draw the graphical symbol for a 4/3-DC single solenoid hydraulic valve.
10. What is the function of an NO type pushbutton? Give an application of the pushbutton.
11. What is the function of an NC type pushbutton? Give an application of the pushbutton.
12. Distinguish NO type and NC type pushbuttons.
13. What is a pushbutton station? Explain with a proper symbol.
14. What is an electromagnetic relay?
15. Explain the operation of an electrical relay with a suitable sketch.
16. What are logic functions, as used in electro-hydraulic circuits?
17. State one application each of 'AND' and 'OR' logic functions in hydraulic systems
18. What is a memory function? Explain two methods of realizing memory function in electro-hydraulic circuits.
19. What are sensors? Explain their importance in electro-hydraulic circuits.
20. Explain the operating principle of a limit switch.
21. What is the area of application of limit switches?
22. What are proximity sensors?
23. What are the advantages of proximity sensors as compared to limit switches?
24. State some applications of proximity sensors.
25. Explain the working principle of an inductive proximity sensor with a block diagram.
26. Explain the working principle of a capacitive proximity sensor with a block diagram.
27. What are photoelectric sensors?
28. Explain the working principle of an optical proximity sensor with a block diagram.
29. Explain the operation of an on-delay timer with a timing diagram.
30. Describe the functioning of an off-delay timer with a timing diagram.
31. Develop an electro-hydraulic circuit for the control of a single-acting cylinder used for the cutting operation using a fixed-volume pump and a single solenoid valve. Incorporate the necessary safety measures in the circuit.
32. Develop an electro-hydraulic circuit for the control of a double-acting-acting cylinder used for the closing and opening operation of a carriage using a fixed-volume pump and a double solenoid valve. The circuit should respond to momentary signals from pushbuttons. The forward and return motions of the cylinder carrying the resistive load must be slow. The pump should unload when the motions reach their extreme end-positions, at the same time should latch the position of the load hydraulically. Incorporate the necessary safety measures in the circuit.
33. Develop an electro-hydraulic circuit for the control of a double-acting cylinder used for a machine tool operation using a fixed-volume pump and a double solenoid valve. The cylinder should return automatically upon reaching its extreme end-potion. Incorporate the necessary safety measures in the circuit.
34. Develop an electro-hydraulic circuit for the control of a double-acting cylinder used for a hydraulic press using a fixed-volume pump and a double solenoid valve. The cylinder should return automatically after 5 seconds, upon reaching its extreme end-potion. Incorporate the necessary safety measures in the circuit.
35. Develop an electro-hydraulic circuit for a continuous cycle of operation for a hydraulic cylinder in a heavy industry using limit switches and relays.

Appendix 1 | Graphic Symbols for Electrical Components

A list of the essential graphic symbols for electrical components is given below.

Symbol	Description	Symbol	Description
	Normally open contact (NO)		General manual switch contact
	Normally closed contact (NC)		Relay contact
	Change-over conttact		Pushbutton contact
	Switch with NO contact (not automatically reset)		Pull button contact
	Switch with NC contact (not automatically reset)		Twist switch contact
	Mechanically linked contacts		Roller switch contact
	Normally open contacts, actuated		Delay to operate
	Pressure switch		Delay to reset
	Switch with NO contact manually- actuated by rotating		Proximity switch

Symbol	Description	Symbol	Description
—	Direct current (DC)	⊐□⊏	Relay coil
∼	Alternating current (AC)	⊐⊠⊏	Relay coil with delayed contact operation
≂	AC or DC	⊐■⊏	Relay coil with delayed contact reset
+ −	Positive and negative polarity		Solenoid coil
⏚	Line to earth		Solenoid actuation
⏛	Line to chassis		Solenoid actuation with manual override
	Cell		Double solenoid
or 12 V	Battery		Single solenoid with spring return
24V 0V	Supply and return lines		Solenoid and pilot actuation with manual override
	Fuse		Pneumatic – Electric (PE) converter

Symbol	Name	Symbol	Name
	Resistor		Light emitting diode
	Potentiometer		Photo-transistor
	Inductor		Opto-coupler
	Inductor with core		NPN transistor
	Variable inductor		PNP transistor
	Capacitor		Triac
	Polarised capacitor		Thyristor
	Variable capacitor		Measuring instruments for current, voltage, resistance and power
	Diode		Motors ac and dc
	Zener diode		transformer

References
1. Joji P, Pneumatic controls, Wiley India Pvt Ltd, New Delhi, 2008
2. Joji Parambath, Industrial Hydraulic Systems, Universal Publishers, Boca Raton, USA

Fluid Power Educational Series Books

1. Pneumatic Systems and Circuits -Basic Level (In the SI Units)
2. Industrial Pneumatics -Basic Level (In the English Units)
3. Pneumatic Systems and Circuits -Advanced Level
4. Electro-Pneumatics and Automation
5. Design of Pneumatic Systems (In the SI Units)
6. Design Concepts in Pneumatic Systems (In the English Units)
7. Maintenance, Troubleshooting, and Safety in Pneumatic Systems
8. Industrial Hydraulic Systems and Circuits -Basic Level (In the SI Units)
9. Industrial Hydraulics -Basic Level (In the English Units)
10. Hydraulic Fluids
11. Hydraulic Filters: Construction, Installation Locations, and Specifications
12. Hydraulic Power Packs (In the SI Units)
13. Power Packs in Hydraulic Systems (In the English Units)
14. Hydraulic Cylinders (In the SI Units)
15. Hydraulic Linear Actuators (In the English Units)
16. Hydraulic Motors (In the SI Units)
17. Hydraulic Rotary Actuators (In the English Units)
18. Hydraulic Accumulators and Circuits (In the SI Units)
19. Accumulators in Hydraulic Systems (In the English Units)
20. Hydraulic Pipes, Tubes, and Hoses (In the SI Units)
21. Pipes, Tubes, and Hoses in Hydraulic Systems (In the English Units)
22. Design of Industrial Hydraulic Systems (In the SI Units)
23. Design Concepts in Industrial Hydraulic Systems (In the English Units)
24. Maintenance, Troubleshooting, and Safety in Hydraulic Systems
25. Hydrostatic Transmissions (HSTs) (In the SI Units)
26. Concepts of Hydrostatic Transmissions (In the English Units)
27. Load Sensing Hydraulic Systems (In the SI Units)
28. Concepts of Load Sensing Hydraulic Systems (In the English Units)
29. Electro-hydraulic Proportional Valves
30. Electro-hydraulic Servo Valves
31. Cartridge Valves
32. Electro-hydraulic Systems and Relay Circuits

For more details, please visit: **htpps://jojibooks.com**

www.ingramcontent.com/pod-product-compliance
Lightning Source LLC
Chambersburg PA
CBHW081059240526
45465CB00025B/2752